CHOOSING YOUR BOAT

Also by John Roberts

Fiber Glass Boats: Construction, Repair, and Maintenance

CHOOSING YOUR BOAT

A Practical Guide to Selecting a Sailboat That Fits Both Your Pocketbook and Your Sailing Needs

John Roberts
Maria Mann

Photos by Susan Roberts

W·W·николтон & COMPANY
NEW YORK   LONDON

Copyright © 1986 by John Roberts
ALL RIGHTS RESERVED.

Published simultaneously in Canada by Penguin Books Canada Ltd.,
2801 John Street, Markham, Ontario L3R 1B4.
PRINTED IN THE UNITED STATES OF AMERICA.
FIRST EDITION

The text of this book is composed in Baskerville, with display type set in Palatino. Composition and manufacturing by The Maple-Vail Book Group.

Library of Congress Cataloging-in-Publication Data
Roberts, John Arthur, 1945–
   Choosing your boat.
  1. Sailboats.   I. Mann, Maria. II. Title.
VM351.R63 1986   623.8′223   86-16386
ISBN 0-393-03315-5

W. W. Norton & Company, Inc., 500 Fifth Avenue, New York, N.Y. 10110
W. W. Norton & Company Ltd., 37 Great Russell Street, London WC1B 3NU

1 2 3 4 5 6 7 8 9 0

Contents

Introduction 11
I NARROWING THE FIELD 13

1 The Challenge 15
Merging Dreams with Reality

Today's boat shopper is faced with an almost bewildering variety of sailboats marketed to fit virtually every dream and almost every pocketbook. Even reasonably experienced sailors, however, can make serious errors in selecting a boat for the sailing they will do.

2 First Things First 19
Appraising Your Boating Needs

There are three ways to go about buying a boat: (1) Trust to dumb luck; (2) pay someone else to make decisions for you; and (3) learn enough to make a reasonably informed decision about what boat is best suited for the sailing you plan and the size of your pocketbook. How will you use your boat? Fitting boats into one of five categories depending upon the use they'll see. Honestly evaluating your personal sailing plans and aspirations.

3 Where to Look 26
From the Safety of Your Home

Developing a list of candidate boats from the safety of your living room. The importance of wives and husbands working together on this search. Information that can be gleaned from advertisements and brochures to help fit boats into each of the five basic categories of use. Separating fact from fiction in advertisements and promotional materials.

4 Where to Look   44
    Shows, Showrooms, and Docks

Boat shows are a good place to test the impressions you gained from advertisements and promotional materials, but leave your checkbook and credit cards home. The value of apparently casual conversation at marinas and yacht clubs. Dealer showrooms as a good place for leisurely and thorough looking at specific boats.

5 How to Look   54
    A 15-Minute Checkup

How to check out a boat in 15 minutes—at a boat show, a dealer's dock, or even while enjoying a Sunday sail on a friend's boat—to measure its suitability for each of the five basic categories of use. The importance of practicing the technique to sharpen your skills before beginning a serious effort to narrow the field to the few boats you want to look at more closely.

II FIBER-REINFORCED PLASTIC (FIBER GLASS) BOAT CONSTRUCTION AND DESIGN   75

6 Fiber Glass Boat Construction   77
    A Look beneath the Gloss

How fiber-reinforced plastic (principally fiber glass) boats are built, with emphasis on the suitability of different construction methods and details for the five categories of use introduced in Chapter 2. Materials of construction, solid and cored laminates, interior installations, and construction details.

7 Boat Design   151
    A Look at Form and Function

Design features to be considered in weighing a boat's suitability for the use intended. Displacement, exterior design features ranging from keel-and-rudder configurations to the type of rig, and key interior design features ranging from cabin space and decor to stowage areas—all related to the five basic categories of use.

CONTENTS

III MAKING A DECISION   177

8 The Second Time Around   179
A Closer Look at the Leading Ladies

Measuring candidate boats against your list of priorities. Fundamental priorities for each of the five categories of use. Taking a measure of the dealer.

9 Visiting the Factory   192
A Final Test

Measuring the health of the builder. Assessing the quality of operations. Testing the effectiveness of quality-control systems at the factory. A chance to look beneath the surface at key construction details.

10 Making Your Choice   200
A Job Done Well

It's a buyer's market. Putting it all together to be certain you've settled on a boat that fits your needs. When and where to buy. Retaining control over negotiations when closing the deal.

APPENDIX: YACHT BROKERS   208
Making Them Work for You

Structure of the yacht brokerage industry. How the system works. Commissions. Selecting a broker. Whom does the broker serve—buyer or seller? Contracts and deposits. Storm warnings to watch for.

INDEX   220

List of Tables

Table Number	Page	
1	103	Typical lay-up schedules for solid FRP hulls
2	156	Characterization of boats by displacement/length ratios—1970s and 1980s
3	180	Fundamental priorities for five boat-use categories

Introduction

Deciding what boat to purchase is one of the most difficult decisions a sailor has to make. In part, the difficulty arises from the inevitable need to compromise. In part too, it arises from the fact that buying a sailboat usually is an emotional process. Particularly with fiber glass boats, however, there is a third source of difficulty—the absence of a widely held conventional wisdom to use as a scale for measuring the quality of construction and the suitability of a particular boat for the use one wants to make of it. The result is that people can make serious mistakes—mistakes which, at the least, can be costly in terms of money and, at the worst, costly in terms of personal safety.

The purpose of this book is to help you avoid such mistakes. First, the book provides the kind of information and practical suggestions that should enable anyone purchasing a fiber glass sailboat in the 25- to 45-foot range to make more informed judgments throughout the selection process. Second, it suggests ways to develop a system for screening new and used boats. Such a system makes it possible to cull quickly those boats whose basic design or construction (or both) make them unsuitable for the use contemplated. The same system also helps identify boats deserving a second, more thorough look. Third, the book provides information needed to understand more completely how fiber glass boats are built and then relate various construction and fabrication methods to different kinds of sailing. Fourth, when it is time to go back for that "second, more thorough look," the book suggests the kinds of questions you might ask, ways to get answers to your questions, and a framework for evaluating those answers when you get them.

Throughout all, we recognize that few people will want to follow all of the suggestions contained in the following pages. We have, however, attempted to fill these pages with enough facts, opinion, and ideas that each person will find the kind of practical information and suggestions needed to make buying his or her next boat as satisfying and successful an experience as it can be. Finally, although this book does not advocate any particular design or construction method, it does present an attitude that we hope will be contagious—an unwillingness to accept compromise in safety.

I

Narrowing the Field

One of the problems facing every prospective boat shopper is deciding which boat among the hundreds available will best suit his or her purposes and pocketbook. Insofar as it is impractical to look thoroughly into every available boat, a system is needed to narrow the field to a select few—each of which on the surface appears suitable for the sailing planned.

1

The Challenge
Merging Dreams with Reality

For most of us, a sailboat begins as a dream. A moment at the beach watching sails slip by along the horizon. A walk along a city dock or marina, admiring the variety of boats lying full of promise in their slips. You may already have a boat and be dreaming about something larger. Or perhaps you have no boat, but dream of buying one someday. In either case, the day will come when you decide to convert your dream to reality. It is a decision repeated every year by thousands of dreamers—so many, in fact, that those who dream have helped spawn a large sailboat industry. It is an industry that has developed in barely 25 years to such levels of sophistication in production and merchandising technology that sailboats are now available to fit almost every pocketbook and, it would seem, almost every dream. And so the challenge is yours—to be able to search among the many boats offered for sale to find one that will suit your own particular needs.

On the surface, the job involved in buying a sailboat may not seem much different from that of buying a house. You know approximately how much money you can spend, what accommodations you want, and you have some idea of the style you like. Talking to brokers and visiting "model homes" (boat shows) provide a chance to sample the market. You can go through several houses (boats) that have been spiffied up to make them as attractive as possible. Then, after asking whatever questions you may have about the one or two that interest you most, you can make a decision based upon how the houses (boats) under consideration fit your needs and pocketbook. The process is essentially the same whether the house (boat) you buy is new or used.

It's a comfortable way to go shopping and one the real estate industry and much of the boating industry encourages. It's probably also an adequate way to shop for a house. But it's a way to get into trouble when shopping for a sailboat. Although a boat can be a home, it is not a house. Basic flaws in design or shortcuts taken in the construction of a house are mostly just irritating—a shortage of closet space, for example; higher energy bills from a lack of insulation; repair bills from a leaking shower installation; or never-ending frustration from a wet basement. In contrast, flaws in the design and/or construction of a sailboat can be life-threatening—even during a summer weekend sail on the Chesapeake Bay or Puget Sound.

The scope of problems sailors can run into is illustrated by the experience of a California couple who decided to move up from a 30-foot sloop they had used for several long cruises along the Pacific Coast to a 42-foot ketch they could live aboard and sail to the far reaches of the world. The boat this couple chose was produced by one of the larger boat builders in the U.S. and has been widely promoted in magazine advertisements as a world cruiser. The interior layout is well suited for living aboard at a dock, though less well suited for ocean cruising. In any case, the boat is generally attractive for her type and it is easy to understand her appeal. Unfortunately for this couple, however, their boat proved the old adage about "beauty being only skin deep." Its history was littered with problems from the day of its arrival in their California boatyard. Fortunately, the owners did not try to cross any oceans before the full extent of their problems was discovered.

"Right from the start," they reported, "there were so many things wrong, and so many things missing that we were staggered."[1] Although they had been assured that commissioning would take only three to four weeks, the yard "spent the first

---

1. Our summary of this couple's experience is adapted with permission from a more detailed report published in *The Telltale Compass,* a privately circulated newsletter for yachtsmen.

eight weeks in warranty work alone, simply patching up manufacturing mistakes." Problems included more than a dozen voids in the fiber glass lay-up of the deck; these had to be ground out, filled, and gel coated and the nonskid ridges cut in by hand. All portlights leaked around their frames because the builder had not bedded them properly in sealant; four of the plastic portlights broke when being removed for proper bedding. The ice chest had only one inch of insulation around it. The mast was not rigged correctly. And so it went. Although the list was long, all of the problems appeared repairable—until, nearly a year after the boat had been delivered, a diver hired to clean the bottom reported that the strut supporting the propeller shaft was loose. When the boat was hauled, the owners found not only a loose strut, but also a misshapen hull. At this point, they called in professional marine surveyors to look at their boat. Among the findings were the following:

(1) A major crack in the fiber glass hull on the starboard side, apparently caused by the hull flexing over a hard spot formed by the junction of the cabin sole and the galley structure, with the result that "structural failure to this area was imminent."

(2) More than a half-dozen major deformities in the hull; "due to the large extent of the deformities, a true hull line could not be obtained."

(3) The keel out-of-line by about two inches.

(4) The cap rails "apparently not properly fastened," with the result that they were loose and cracking at some of the fastenings.

(5) A loose chain plate for the mizzenmast.

(6) A serious water leak directly over the master electrical panel.

(7) A large crack in the cabin sole, possibly indicating "excessive working of the vessel."

(8) The boat 5,000 pounds lighter than the 30,000-pound designed displacement, the result either of insufficient ballast, a fiber glass lay-up that was below specifications, or a combination of both.

In addition, a report on core samples taken from the hull showed the starboard sample was fractionally less than a quarter-inch thick. Moreover, the resin-to-glass ratio of the samples was higher than good practice allows. The report noted that "the total weight of glass installed does not satisfy any of the American Bureau of Shipping requirements. For a displacement vessel or for curved panels, the weight of glass installed is grossly inadequate." It said also that the construction method used in the boat "is inadequate for the service intended."

Obviously, no boat builder can hope to stay in business long if he consistently produces such disasters. And, hopefully, the boat just described represented an aberration for the builder involved. However, structural failures can occur in virtually any boat subjected to conditions beyond those for which the boat was *designed and constructed.* Furthermore there are any number of boats on the market today that would not have to be tested very far to bring about structural failures. The simple fact is—despite the claims of many advertisements—that most sailboats sold as cruisers or racer/cruisers today are not built to cross oceans. Most also are not built for coastal cruising if that means getting more than a few miles from a protected harbor. In fact, a surprising number of sailboats are not even built adequately for safe sailing in the semiprotected waters of the lower Chesapeake Bay or San Francisco Bay if the weather acts up a bit. This is not necessarily bad. It does, however, add challenge to the job of choosing the right boat for you—a challenge to learn enough about boats to find one that not only fits your pocketbook, provides the accommodations you want, and fits the image you have in mind, but one also that is designed and constructed for the kind of sailing you want to do. In short, the challenge is to make buying your sailboat an ACTIVE experience in which you buy a boat suitable for you, rather than a passive episode in which you are sold a boat someone else says is what you want.

2

First Things First
Appraising Your Boating Needs

Fiber glass—fiber-reinforced plastic, or FRP as it is more appropriately called—is an excellent boat building system. More than any other single development, it has brought assembly-line technology into boat building and helped bring the price of boats down to a level the majority of us—not just the wealthy—can afford. Moreover, the durability and performance of fiber glass boats has been instrumental in introducing bank financing to the boating industry. As a result, people today are buying larger and larger boats, and almost anyone who wants a boat can buy one as long as his credit rating is healthy.

At the same time, however, the use of fiber glass technology in building boats has made it more difficult for nonexperts—and that is most of us—to distinguish between well-built boats and those that are not so well constructed. As a result, most of us are at a decided disadvantage when setting out to look for a boat if we are concerned about finding a boat well suited to our needs.

There are, however, some options available. For example, we can press ahead blindly as too many people do and trust to good fortune that we will wind up with a boat suitable to our purposes. Another option is to hire the services of a marine surveyor to screen our choices for us and then do a complete survey on the final one or two contenders. Still another alternative, however, is to learn enough about boats to do our own screening, narrow the field down to a final candidate, and then employ a professional surveyor for a check on our judgment if we feel it necessary.

Our own preference is for the third option. It is far safer than the first. It costs less money than the second. It is much more fun than either of the first two. And it's more satisfying too. It feels good, for example, to recognize the problems you have saved yourself as you go through a boat you once thought about buying and begin to notice shortcomings in design or construction that make the boat unsuitable to your needs. Moreover, when it comes to selecting the specific boat you want and making the inevitable compromises, the effort you have put into getting that far in the selection process makes compromising easier. The reason is that you know more fully what the compromises are and their implications for your sailing.

There is still another advantage to learning enough to make your own decisions: That is, you gain a measure of confidence in yourself and in your boat that you cannot purchase from someone else; that confidence just adds to your sailing pleasure.

So where do you start?

You begin by making two commitments to yourself. Commitment No. 1 is that you are not going to rush into a purchase. As with many other things that are worth doing, if it is done well, shopping for your boat will take time. At the least, that time probably will be counted in months; it is not unusual for it to run well over a year; and, for some people (including one of the authors), the process may require three or four years.

Taking this amount of time to shop for a boat can be a problem. By the time you get to the decision to start looking for your boat, you probably want it NOW! There are, however, ways to keep that yearning in check. One is to start the search before you are in a position financially to buy the boat. With some luck and planning, your selection process and financial situation will mature at about the same time. Assuming you have already planned the financial part of the purchase, there are more creative ways to resist the urge to buy now. If you already own a boat and now want to move to something else, you have an advantage. You can plan on an extra season or so with your present boat and perhaps get a new sail to spice up sailing for the added year. If you do not already have a boat, you might consider taking the money it would cost in upkeep, etc., for the next year if you bought a boat immediately and spend that money instead on a sailing school or bareboat chartering, either in local waters, the Bahama Islands, or the Caribbean. Often, you can charter boats of the type you are considering and, in that way, see if the boat lives up to the builder's claims. Or, if you are moving to a larger boat, the cruising schools and bareboat charters provide opportunity to gain experience sailing larger boats—before you purchase yours.

If you do not now have a boat, a second option is purchasing a day sailer that can be kept on a trailer. The upkeep is minimal. The sailing is fun. And it is a chance to hone basic sailing skills while you look for the larger boat. In addition, as long as your day sailer is a popular boat in your area, it probably can be sold after a year or two for very nearly what you paid for it.

So plan on putting your boat kitty into certificates of deposit or a good money-market fund, use the estimated cost of the boat you will not have this first year to take the edge off your impatience, and make Commitment No. 2—that you are going to use the next several months to learn as much as you can to help yourself make a good decision in selecting your boat. In that way, when you finally do get the boat of your dreams, you can be confident that your dream will not turn into a nightmare.

HOW MUCH BOAT DO YOU WANT?

Every cruising sailboat will see foul weather someday. If a boat never leaves the dock, perhaps all the skipper has to be concerned about is having strong mooring cleats and adequate ventilation even when all ports and hatches are closed. However, if a boat is taken for a day's sail into a large body of water and unexpected changes in wind direction and strength confront the skipper with a 15-mile beat into 25- to 30-knot winds to reach a safe harbor, that skipper could have a problem if the boat is not up to handling those conditions.

Because there are great differences in the demands placed on a boat depending upon the conditions it encounters, we find it useful to fit cruising sailboats into one of five categories. These categories are based on the areas in which the boats will be used.

(1) Dockside—This boat will rarely leave the dock. It is principally a floating second (or even a first) home. When it does leave the dock, this boat usually should be kept in protected or semiprotected waters.

(2) Protected waters—This boat is best restricted to use on smaller sounds and rivers. Examples include Albemarle Sound, Pamlico Sound, Biscayne Bay, the upper reaches of the Chesapeake Bay, small and midsize rivers, and smaller lakes. In general, land is always close-by and there is not enough open water for heavy waves to build up before the boat can reach shelter.

(3) Semiprotected waters—This boat can be used safely in large bays and sounds. Examples include Long Island Sound, Chesapeake Bay, Puget Sound, and San Francisco Bay. These waters can get quite rough very quickly, but there are numerous harbors and lots of help nearby if needed.

(4) Coastal and nearby offshore waters—This boat is generally suitable for coastwise sailing in the ocean under good weather conditions and with the expectation of going into port each night. With proper care in choosing favorable weather and sea conditions, it would also be suitable for a trip across the Gulf Stream to the Bahamas, a hop out to Catalina, or an overnight Down East from Cape Cod to the Maine coast. With careful preparation and an experienced crew, this boat may also be suitable for an occasional longer passage—for example, a trip to Bermuda.

(5) Ocean passages—This boat is intended for crossing oceans. It is designed, constructed, and equipped to be independent of land for periods of a month or longer.

To a degree unsuspected by most people, these five categories describe such details as design features, interior arrangements, ruggedness of construction, and—often—price. That's why it is important to decide at the outset where you will be doing your sailing. It also helps to be realistic. For example, all of us may dream of sailing out to the islands of the South Pacific; most of us, however, will not begin by crossing oceans. For this reason, it usually—not always, but usually—makes sense to buy a boat for the sailing you will be doing during the next few years rather than one suited for the sailing you hope to be doing ten years from now.

It helps to be a bit practical as well. Purchasing a true world cruiser to camp out weekends at the yacht club is a costly ego trip, especially when there are a number of other "cruising" sailboats much better suited to dockside weekending and probably available at a lower price—sometimes for even half the price! Similarly, it makes little sense to buy a boat suitable for sailing to Bermuda if all you have is a two-week vacation and will be keeping the boat in Long Island Sound. With a four-week vacation and bit of forward planning, the boat for a Bermuda cruise may make more sense. On the other hand, if coastal cruising or trips to Bermuda are your kind of sailing, it would be a false economy—not to mention a serious gamble with your life—to buy a boat intended only for use in semiprotected waters. Assuming the boat would not sink during a severe gale offshore, it would be a constant source of headaches and expense for repairs because it had not been built ruggedly enough to hold up under the use it was receiving. Two examples illustrate the point: One stock boat we know that was sailed in a Bermuda race had one of its key bulkheads come loose during a stormy return from Bermuda—after the race was completed. Another stock boat raced successfully in both Bermuda and OSTAR races has had bulkheads come loose twice! The first time it reportedly cost $17,000 to have them bonded back to the hull. We do not know about the second time, but the moral should be clear: There is no sense in overbuying, but it is not safe to underbuy. When the time comes that you are ready to move to either more demanding or less demanding waters, there will be ample opportunity to get a boat more suitable to the new waters you will be sailing.

While you are thinking about the use your boat will receive, there are also some personal factors you need to consider—particularly if you are married. The question of children is a good place to start, if you have them. You need to look clearly at the question of whether they will be sailing with you. Many parents have discovered only after getting a 40-footer to accommodate all the youngsters that the kids have other ideas about how they want to spend their weekends and the "family" vacation. In one case we know of, realization that the children would soon be going off to build lives for themselves led a couple to scale their aspirations down from the 42- to 45-foot range to 36 to 38 feet—a huge difference not only in initial purchase price, but also in ease of handling and in maintenance and operating expense.

Not only do couples need to consider carefully what their children's desires are, they also need to look honestly at their own and then to discuss them candidly with each other. Any experienced cruising person has seen sailing dreams go up in air and marriages come apart because couples had not faced one fundamental set of questions squarely: "What do we together expect from our sailing plans?" Not, what does the husband expect? Or, will the wife give it a try? But "what does the husband expect? What does the wife expect? And, what is a good compromise that will give both husband and wife a chance to enjoy the boat they wind up purchasing?"

You also need to consider your age and sailing experience. How much boat can you and your wife together—in an emergency, you or your wife alone—really handle? If you are in your thirties and have been sailing on displacement boats for a number of years, the answer may be quite different than if you are about to start collecting social security retirement benefits and have just started learning to sail. We are deliberately picking extremes of age and experience, but only to make a point. Obviously age alone—assuming reasonable health—does not preclude any cruising you may want to plan. If your age is an important consideration, it is just something you factor into your decisions. In the same way, you also need to factor in your sailing experience.

A picture should now be emerging from your answers to these questions. Perhaps it is a picture of a boat for week-ending in semiprotected waters plus an annual two-week vacation cruise in the same body of water. The children are small and you want them to learn to enjoy sailing, so you want a boat large enough to accommodate the four of you comfortably. Your marriage partner has been sailing just a few times and does not know how to sail very well. You sailed a lot in a daysailer as a teenager and have been a guest a few times on a friend's 35-footer, but truthfully your experience with a cruising sailboat is limited. You think 35 feet is too big for now. . . . Or perhaps you and your partner are ready to retire to a boat and you want to winter in the south and summer in the north. . . . Or, you want to take two or three years to explore other parts of the world from your sailboat. Whatever the picture, you are on your way.

3

Where to Look
From the Safety of Your Home

Once you have settled on the kind of sailing you want to do and the number of people your boat has to accommodate, it is time to begin the search. The goal at this stage is to work up a list of a dozen or so boats within your size range—for example, from 29 to 32 feet—that appear on the surface to be suited for your needs. Developing a list of twelve may mean reviewing two or three times that number of boats. The key point at this time, however, is to accomplish most of the work from the relative security of your living room so that you are not tempted to place an order by a persuasive salesman.

The resources available without straying farther from home than your mailbox or the local library are greater than you may realize. They include boating magazines, promotional materials from boat builders, and books. If you are male and married, however, we offer one emphatic suggestion before you look at the first magazine or brochure: Encourage your wife's participation in this search from the outset and make whatever effort is necessary to keep her involved throughout. The reasons are very practical. First, it will add a positive dimension to her enjoyment of the boat when you get it because she will have been an informed and active participant in selecting it. Second, the more your wife knows about boats generally and your boat specifically, the better equipped she is as your sailing partner. And third, sailboat advertising, brochures, and sales pitches all are aimed to a large extent at women.

The rationale for this advertising strategy is the builder's belief that when the time to make a decision arrives, the woman of the family has important influence on which boat is finally selected. There is, of course, at least some truth in that notion. Few men, for example, will knowingly select a boat their wives do not like—at least, not if they want their wives to sail with them. Moreover, many husbands whose wives are not sailors make compromises against their better judgment in order to keep their wives happy.

A wiser course is to help your wife learn enough to make informed judgments right along with you as the two of you seek the dream you'll be sharing. To understand what you are up against, consider a boat's interior. Boat builders have found it is easier to sell boats that have a familiar look about them. Insofar as a sailboat pretty much has to look like a sailboat on the outside, there is only one place left to develop that "familiar" look—particularly for an inexperienced sailor—and that is on the inside. As a result, the interiors of many boats look as if they have come from the pages of *Better Homes & Gardens*. For some limited uses, such interiors are fine. For much of sailing, however, they are grossly impractical; they may also be unsafe. Yet this is the area of the boat focused on by most of the advertising, promotional brochures, and sales strategy aimed at women. The only protection you and your wife have against this sales strategy is to learn at the outset about the pros and cons of different interior features for the kind of sailing you plan to do and then screen the boats accordingly.

MAGAZINES AND NEWSPAPERS

Boating magazines are a rich source of information. The chances are good that you already purchase at least one of them. Strong argument can be made for subscribing to three or four—perhaps even five if you want to include one of the lesser known publications—while you are trying to select your boat. It may cost you $70 or more to subscribe to the four majors—*Sail, Yachting, Cruising World,* and *Motor Boating & Sailing*. That may seem like a lot of money for magazine subscriptions, but it's a small investment when placed next to the $25,000 to $250,000 or more you plan to spend for your boat. By the time those first-year subscriptions are up for renewal, you'll probably not need the magazines any more to help with your search for a boat, but you may want one or more of them to keep up with advances in equipment and the news of boating. If so, you will have had ample opportunity to decide which of these magazines will best serve your continuing need.

Two other publications also should be considered. The first is *The Practical Sailor;* the second is *Soundings. The Practical Sailor* is the sailing industry's version of *Consumer Reports*—but on a much smaller scale. The publication is issued every two weeks. It contains no advertising; it is wholly supported by its rather steep subscription fee of about $50 per year. It normally has from eight to sixteen pages and is punched for a three-ring binder. While much of the page space is devoted to reviews of equipment, two recurring features are particularly useful to anyone looking for a boat. One is periodic discussion about boat design and construction as they relate to seaworthiness. The other is periodic reviews of boats. Both features are instructive.

If your search is leading you to a used boat, *Soundings* may be of particular value once you are ready to buy your boat. This monthly tabloid newspaper is sold both by subscription and over-the-counter at marinas and some newsstands in states bordering the Great Lakes and the East, Gulf and West coasts. The *Soundings* brokerage section is an excellent supplementary source of information about the used boat market. It includes many private listings as well as brokerage listings from marinas too small to justify advertising in the major magazines. Along with the classified ads in your local Sunday newspaper, *Soundings* will be an excellent source of information about the used boat market in your region. It is worth noting, however, that most of the more successful boats produced in the recent past are still in production today. As a result, the used boat buyer has much to gain by starting out using the same resources and following much the same path as someone looking for a new boat.

Reading Advertisements

Although the advertisements in boating magazines are an obvious first place to start looking for your boat, you will probably need to look at those advertisements differently than you have in the past. For example, you need to recognize at the outset that advertisements usually are not intended to provide much substantive information. The major purpose of boat advertising is to attract reader interest by creating a good feeling about the boat and the people who build or sell it. Your major purpose, however, is to extract facts. This means you need to look past the vase of flowers and bowl of fruit placed so carefully in the photo; for most of sailing, vases of flowers and bowls of fruit are unrealistic anyway. You'll also need to read right over such claims as "liveaboard comfort for six" in a 36-foot sailboat (not really possible) and look instead for clues—positive or negative—as to how the boat is really intended to be used.

Generally, you will learn more about a boat by looking at the photos in the advertisements than by reading the words. For example, even though the text of an advertisement may speak of "coastal cruising," a photo showing a boat with single lifelines should make you wonder how serious the builder is about his boat going offshore. Similarly, if the photo shows neither lifelines nor a stern rail and bow pulpit—as at least one advertisement for an "ocean cruising boat" has done—you have to wonder whether that builder has any idea what ocean sailing is all about. Certainly any sailboat intended for coastal or offshore use should have sturdy double lifelines and bow and stern rails.

There are other visual clues as well: large windows on boats under about 45 feet tend to mean mainly semiprotected waters or less because they can easily be broken by the water in heavy seas; such windows generally do not mean ocean cruising. Large, open main cabins that look like a living room tend to mean dockside living and sailing in protected waters. In a seaway, people can be thrown across a cabin easily; a smaller cabin usually offers more frequent handholds and shorter distances to fall. Large aft cockpits also suggest semiprotected or protected waters; the larger cockpit may threaten the boat's stability if it is filled by a breaking wave. However, these are only clues. It is a good idea to note them, but you will need much more information. And, once again, the advertisements can be helpful.

Every advertisement for a boat includes an address to which you can write for more information. Often, if a company builds more than one boat, the advertisement will list the other models so that even if you are not interested in the specific boat featured, you can seek information about one of the other models. If you want information about boats from several different companies, you can use the reader service post cards in the back of some magazines. The magazine's advertising department will forward your requests to the builders for you. It is a service they provide to show advertisers the magazine's effectiveness in attracting inquiries.

Sometimes an advertisement will specify a small charge to receive promotional materials. That is an attempt to keep a lid on frivolous inquiries; all of those brochures cost money to produce and send through the mail. If you do not want to spend any money for the promotional materials at this point in your search, do not. But do write a letter explaining that you are just beginning to look for a boat in the X-foot range and would appreciate receiving whatever materials they can send to help you determine whether their XYZ-35 (or whatever) is one you should consider more carefully. In most instances, you will get a good response.

Boat Reviews

Advertisements, of course, are not the only resource these magazines have to offer. In fact, it is the other resources that justify getting as many of these boating magazines as you can reasonably afford. For example, *Yachting* and *Motorboating & Sailing* traditionally have published extensive reviews of different boats in each issue. Other magazines also publish reviews, but they have tended to be more superficial.

You can wonder whether magazines that rely on advertising revenue from the boat builders would ever say anything negative about a boat. In that question lies the origins of *The Practical Sailor*. Indeed, the boat reviews in *The Practical Sailor* tend to be more constructive than those in the glossy magazines. However, each of the magazines handles the potential conflict of interest between boat reviews and advertising in its own way. At *Yatching,* for example, policy in the past has been that they will not publish staff reviews of boats they think are poorly built or designed. That policy reportedly cost the magazine the loss of advertising from at least one large boat builder for several years.

The problem may be handled in other ways as well. For example, for a while several years ago. *Motorboating & Sailing* had Hal Roth review boats for its readers. Roth—who, with his wife Margaret, is widely known from his books and articles about their many miles of ocean cruising—generally managed to point out positive and negative features about the boats reviewed. When construction details which might be open to question were involved, however, Roth walked a very careful line between offering his own opinion and reporting the builder's viewpoint. For the reader, or course, it was Roth's opinions that were important—as in a review of a series of pilot house sailboats, when he commented at the end of the article: "In our rush to promote this fine new class of yachts, let's not forget that *they are primarily intended for sheltered waters"* (emphasis added). That is the kind of statement the reader needs to notice—22 words out of an article roughly 1,500 words long.

A one-year subscription, of course, brings only twelve reviews from each magazine. However, a short letter sent to the editors requesting a list of the boats they have reviewed over the past year or two and the issue dates of each review offers you access to the recent past. If you do ask for such a list, be sure to send a stamped, self-addressed envelope and ask about charges for obtaining back issues. Depending upon how many of the boats on the magazine's past review list are among those you are considering at this point, you may want to check your local library to see if they have back issues. You may also want to narrow your selection a bit before buying up back issues if your library does not stock them.

Cruising World traditionally has offered three interesting resources. One is their "Sailboat Show Annual." At this writing, that special issue contains illustrations, layouts, and descriptions of a hundred or more boats, most of them in the 25- to 45-foot range. It is a handy reference.

The magazine also operates a service it calls "Another Opinion." In essence, this service gives boat shoppers a way of "asking the guy who owns one." Boat owners register their names, addresses, and boat type with the magazine, which then publishes a list each month of all the different boats registered—a list which now contains several hundred different boats. You simply send the magazine a stamped, self-addressed envelope along with a short note listing the boats you are interested in and they send you a list of people who own those boats. It is then up to you to contact those people for their opinions.

Brokerage Advertisements

The brokerage section of boating magazines is another good source of information. Once you start accumulating new boat price information, for example, the brokerage ads give you an opportunity to compare how well different boats have tended to hold their values. Until recently, most production sailboats have tended to sell on the used market for at least what they cost when originally purchased. Many are selling for a price that is higher than their original cost because of inflation. A few have even closely tracked the new boat price tag as it has gone up, staying only 15 percent or so behind the new boat replacement cost for at least the first few years. Viewed within this framework, resale value offers a reasonable barometer of value.

You should not, however, try to compare the price asked for an individual 1980 XYZ-40 to the cost of a new XYZ-40 today. The price comparison may be skewed if that particular used boat includes an unusual amount of extra equipment in the asking price. You can eliminate much of the equipment variation, however, by averaging the asking prices of several 1980s and then making the comparison. In this way, you are comparing the asking prices, for example, for an "average" 1980 Out Island 41, an "average" 1980 Irwin 42, and an "average" 1980 Valiant 40 to the new boat prices of each. In this way, you will wind up with a reasonably good idea of how well each boat has held—or increased—its value.

Another clue to boat quality can be found in the numbers of any one boat model listed for resale. If you are looking at the brokerage ads late in the year, for example, and find ads for two or three of that same year's XYZ-35s, plus several listings for XYZ-35s from the last year or two, then you have to wonder why so many people want to sell their XYZ-35s so fast—at least you do if that is one of the boats you are considering. On the other hand, if a builder tells you he has sold 30 ABC-33s a year for the past six years, and you find very few listed in the brokerage sections, that suggests owners of the boat are happy with it. However, you may want to check further in your local newspaper and in the *Soundings* brokerage section before drawing that conclusion—particularly if it is a smaller boat. Many of the brokers who advertise in the major national magazines refuse listings for smaller boats. There is not enough profit in them.

PROMOTIONAL MATERIALS

When you begin requesting information from boat companies, prepare to receive a varied collection of brochures. Some are carefully designed, expensive booklets filled with color photographs. Others are poor quality photocopies of typewritten sheets. Although there is a natural temptation to make generalized assumptions based on the appearance and content of these brochures, it is a temptation to be avoided. Yes, it would be nice to assume that builders who send professionally created color brochures about their boats are high-quality builders whose approach to boat building is accurately reflected in the quality of their brochures. Similarly, it would be convenient to assume that those who provide detailed descriptions of how well their boats are constructed are indeed selling well constructed boats. That, however, would make your job too easy. And besides, appearances too often are deceiving. All that an attractive brochure means is that the builder has the marketing sophistication and money to hire a competent advertising agency to produce his promotional materials. Similarly, all that a lot of well-chosen words about how a boat is put together demonstrates is that the builder (or the advertising copy writer) knows how to talk a good game. What all of this boils down to is that promotional brochures are much like advertisements: Usually you have to dig for the facts you need.

As is the case with advertisements, photos can be helpful. For example, you may see in the photos that the companionway opening reaches almost to the cockpit sole—O.K. for protected and semiprotected waters; questionable for coastal cruising; and a definite negative for crossing oceans. A wave breaking into the cockpit would just flood the boat if the companionway were open.

You can also look in the photos for handholds above and below the deck—a necessity for safe sailing in any but placid waters and light air. Photos can also reveal rounded or sharp corners both above and below deck. Sharp corners on the sailboat are weapons waiting for you to fall on them. Cockpit size, mentioned earlier, is an important clue. Racing boats may justify a large cockpit because of their large crews, but even then the transom often is open to the sea so that the cockpit can drain quickly. On boats seriously intended for coastal or offshore sailing, small cockpits with large cockpit drains are preferred.

Photos of the interior will usually show the galley sink location—another important clue. An outboard sink is fine for use at anchor or at a dock, but not so fine for any serious sailing. A sink should be located as close as possible to the boat's fore-and-aft centerline if it is to drain well when the boat is heeled. Even more importantly, an outboard sink may allow sea water to siphon back into the boat when you are heeled well over and the sink is on the lee (low) side (Fig. 1). If the weather were bad, the boat closed up, and the crew on deck, serious flooding could result before anyone realized there was a problem. Although it is a good practice to keep sea cocks closed except when there is need to have them open, having the galley sink on the centerline helps protect against human forgetfulness—i.e., an open sea cock. Moreover, what is good for the galley sink is good also for head sinks. They too should be mounted along the centerline or close to it.

Having fiddles high enough to keep things on shelves, countertops, and tables is another feature you can look for in photos

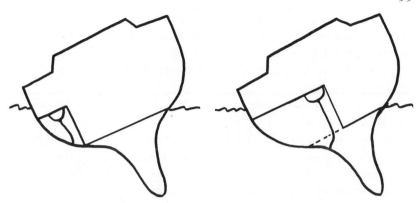

FIG. 1. An outboard sink (left) can easily be put below the water level, letting sea water flood into the boat through the sink unless the sink drain sea cock is closed. Under the same conditions, a sink near the centerline (right) remains well above the outside water level.

to help you categorize the boats. You will want two- or three-inch fiddles on a boat for ocean sailing whereas one-inchers may be adequate on a boat for protected or semiprotected waters. A boat without any fiddles may be acceptable for dockside living, but will prove an exercise in frustration almost from the time the boat leaves the dock and the first sail is raised as one thing after another falls off shelves, counters, and table.

Another clue you can look for: A top-opening icebox or refrigerator is almost a must for any but dockside living. Side- or front-opening boxes not only spill out the cold, but their handles are a shin hazard. Opening a side- or front-opening icebox during or after a day's sail can also result in an "avalanche." The icebox's contents may have been piled up against the door by the boat's motion.

Drawings in promotional materials intended to help illustrate a boat's interior layout also may be helpful. If the photos do not show where the sinks are located relative to the centerline, the drawings will. If photos cannot be used to help determine what berths will be useful while under sail—i.e., how many sea berths there are—the drawings can. A sea berth is one you can lie in securely in a rough sea. Generally, sea berths include pilot berths, single quarter berths ("double" berths are too wide), and the leeward settee. A boat promoted as a family boat for ocean cruising, for example, which relies on a single settee as a sea berth (or none if you are on the wrong tack) should make you wonder either how much thought has gone into the design or whether the boat is suited for the use for which it is being promoted, or both.

Sometimes these drawings or the text of a brochure will detail the location of water and fuel tanks. The preferred location is under the cabin sole to help keep the weight low in the boat. One hundred twenty-five gallons of fuel and water, for example, weighs almost 1,000 pounds and may account for a significant portion of a boat's total designed displacement. Tanks that take up space under settees greatly reduce stowage space—perhaps not such an important consideration for a one- or two-week cruise, but more so if you are considering crossing an ocean. A water tank under a vee berth forward not only reduces stowage space, it also puts weight in the bow, where it may increase the boat's pitching motion.

Stepping back from detail, drawings and photos also provide opportunity to visualize the overall interior to see whether it might work for you—not only in terms of how many crew members it will accommodate, but also in terms of the way the boat is laid out for the kind of cruising you want to do. This too will help you categorize each boat. There is an important point to be aware of, however, in looking at drawings: The drawings illustrate the maximum beam of the boat. Often, because of the curve of the hull, the actual usable width of the boat is less than the drawings suggest.

You will also want to look in the brochures and fliers for whatever specifications are listed for each boat. Recognize at the outset, however, that you are looking for two different sets of specifications—*design parameters* that tell you such vital data as the overall length (LOA), length on deck (LOD), load waterline length (LWL), beam, draft, displacement, sail area, engine size, and tankage for fuel and water; and *construction specifications* that tell you what materials are used to build the boat.

Many builders volunteer only design specifications and then retreat behind a barrage of rhetoric to tell how well the boat is built. Others provide design and some construction specifications, but then still burden you with rhetoric. In any case, you are challenged to separate fact from fiction. For example, the statement found in some brochures, "meets or exceeds safety standards" is fiction—pure rhetoric because it does not identify whose safety standards are being met. The same comment can be made when the Smith boat company says its XYZ-37 is built to "Smith's specs" (whatever they are!), or when the boat is said to be built to "industry standards" (there are no industry standards).

Even when a recognizable set of standards is identified, the statement may be questionable at best, misleading at worst. There is an important difference, for example, between the following two statements, each of which can be found in some brochures: "Hull is built to Lloyds of London specifications"; and, "Lloyds certificate available." The first is merely a claim. Moreover, whether the claim is true is irrelevant because it looks at the hull in isolation from the other components of the structure. The second is a statement of fact. The builder will supply independent certification by Lloyds Register of Shipping that the boat is constructed to a set of specifications recognized worldwide.

If you are confronted by rhetoric about a boat's construction, and if that boat is of interest to you at this stage, a specific request to the builder for detailed information about the hull and deck lay-up schedules, the bulkhead thicknesses and bonding schedules (or details of any other method of attachment), and the hull-to-deck joint will probably yield that information. If the builder will not send you that information, we suggest you strike that boat from your list. One further note: A statement of how thick the hull or deck is without a description of the lay-up schedule and a drawing showing location of reinforcing stringers, frames, and bulkheads is of little real value. The important information, as we will discuss in the second section of this book, is how many layers of what kind and weight of reinforcing fabric are in the lay-up at different points around the boat, what kind (if any) of core materials have been used in the hull and deck, and where and what kind of interior frame members have been used to stiffen the hull and deck structures.

In any case, the idea is to start comparing specifications for different boats by listing them to see what you find: LOA, LOD,

LWL, beam, draft, displacement, ballast amount and material, sail area, lay-up schedule, core material in the hull and /or deck, bulkhead attachment, hull-to-deck joint, chain plate attachment, keel type, water and fuel capacities, engine size and make, total berths, sea berths, sink locations, cockpit and window sizes, etc.

As you look at these pieces of information, you will start seeing more clues to help you place each boat in its appropriate category or, in some instances, categories. For example, a 45-foot boat with only three layers of woven roving and several layers of mat throughout most of the hull probably is not a boat for serious ocean cruising, but may be well suited for dockside living or for sailing in protected or semiprotected waters.

Alternatively, you may begin noting differences between displacement and sail area from one boat to another. A heavy displacement boat, for example, will not sail as well in light air as a lighter displacement boat with a similar waterline length—unless it has a relatively large sail area. This is the reason for the so-called "tall rig" offered as an option on some moderate-to-heavy displacement cruising boats. There are two clear penalties with the tall rigs, however: First, you have to pay for larger sails and for the heavier standing rigging, running rigging, and winches needed for those larger sails. Second is a grey area; one might reasonably ask the designer—not the builder—whether the hull, deck, and chain plate structures designed originally for the shorter rig are adequate for the tall rig without some beefing up. If the designer says beefing up is recommended, the obvious next question (Has it been beefed up?) is for the builder. All of this relates to how you plan to use your boat, and in what waters.

Displacement also gives another clue to the use intended for each boat. Displacement is closely related to interior volume, particularly to the volume below the waterline. From a practical viewpoint, this means that a moderate-to-heavy displacement boat usually has more room low in the hull for the engine, tankage, bilge sump, and stowage—not so important for sailing in protected or semiprotected waters, perhaps, but progressively more important for the needs of coastal cruising, dockside living, and crossing oceans (Fig. 2).

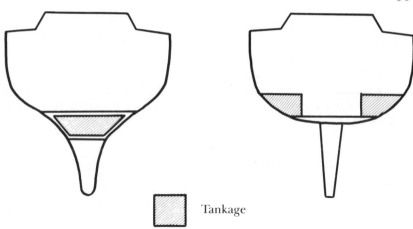

Fig. 2. The larger volume beneath the cabin sole of a heavier displacement design (left) can be used for water and fuel tanks, keeping the weight low and along the centerline. In light displacement designs (right), tankage must be placed above the cabin sole—typically in the settee bases.

Of course, you cannot ignore price. However, you need to develop a list of "standard" features and equipment to provide a standard basis for comparison. In that way, you can adjust the price of individual boats up or down to bring them to that standard. Eventually, too, you will need to consider the quality of construction in making any price comparison. It is often possible to evaluate the relative quality of construction of different boats on a preliminary basis even without seeing the boats firsthand. Leaving out for the moment any discussion of exotic materials of construction, and assuming we're talking about boats made in the United States, certain assumptions can be made. For example, one large builder tends to pay about the same price for materials of construction as any other large builder. Smaller builders may pay slighter higher prices than larger builders, but all of the smaller builders are in the same market. Labor costs also are often similar from one builder to another, though there may be some regional variations depending upon the labor pool in each area. What these facts add up to is this: Price often reflects quality of construction.

To relate price to quality, however, you must first convert the price for each boat (after adjustment to a standard list of features and equipment) to a price per pound of displacement. For example,

> The ABC-36's design displacement is 18,000 pounds. The selling price (adjusted to standard basis) is $120,000. The price per pound of displacement is:
>
> $$\frac{\$120,000}{18,000} = \$6.67 / lb.$$
>
> The XYZ-37's design displacement is 20,000 pounds. The selling price (adjusted to standard basis) is $89,000. The price per pound of displacement is:
>
> $$\frac{\$89,000}{20,000} = \$4.45 / lb.$$

Looking at the data you are organizing, you should see a trend which shows boat prices (expressed in dollars per pound) getting higher as the demands of the sailing for which the boats are suited get tougher. Significant exceptions to such a trend deserve additional questioning to be sure you have categorized the boat correctly if it is one of the group you are considering. Similar comparisons can be made for boats built in other countries; however, it would be difficult to draw any conclusions related to quality of construction by comparing the price per pound of displacement of boats built in different countries because of the different cost structures as you move from one country to another.

BOOKS

Although magazines and promotional materials offer a way to survey the market from the safety of your living room, you will have to turn to books for the kind of information required to evaluate what you find in your survey. How much you need to read depends upon what you know already and your cruising aspirations. At the least challenging end of the spectrum—dockside living—you probably do not have to read a great deal merely to find a boat suitable to your needs. However, the more you read, the better equipped you'll be to evaluate the variety of available options. Also, you need to read and study enough to recognize the limits of whatever boat you choose. Today you may plan only on dockside living; tomorrow you may start thinking about a dock in Bermuda. Or Hawaii. If you have learned enough to know the limitations of your boat, you'll know whether you can plan the trip safely with that same boat, you can modify the boat to make it more suitable for the trip, or whether you should replace it with one better suited for ocean sailing.

At the other end of the spectrum—crossing oceans—you cannot read too much. Between the two extremes, you will have to use your own judgment. Our advice in this as in almost every other aspect of sailing, however, is to err on the side of safety. In short, read more than you think is necessary. Sailing, boat design, and boat construction are inexact sciences at best. As a result, there are almost as many shades of opinion about these subjects as there are sailors. This means that you need to know enough about the range of opinion to be able to form your own with some confidence in its validity.

One book you may well find useful is *Practical Boat Buying*, published by the editors of *The Practical Sailor*. The book is essentially a compilation of articles and boat reviews from *The Practical Sailor* during its first four years of publication. The first 95 pages cover a variety of related subjects ranging from tips for looking for a boat, to the economics of buying a boat, to the cost of keeping a boat after it's been purchased. The next 125 pages consist of reviews the editors made of thirty-two boats ranging in size from 22 to 42 feet LOA. Frequently the reviews are blunt, but they are always interesting. Given the nature of the twice-monthly parent publication, the odds seem good that this book will be updated periodically with additional reviews.

Perhaps the best general source of books about boating is the Dolphin Book Club, a division of the Book-of-the-Month Club. The club offers significant bargains to encourage initial enrollment. As important, the booklets sent out each month to describe the monthly selections provide a current catalogue of what is available through your local library. Although many of these books may not be carried on your local library's shelves unless you live in a major sailing center, all should be available through interlibrary loans—a service available to library members.

Five books you may want to consider purchasing and reading several times over—particularly if you are interested in any ocean sailing—are listed below. Although these books all were written (or revised) several years ago, they offer a range of viewpoint and commentary that remains valid today and should be helpful in developing an idea of what you want in your boat. Their authors all are experienced sailors.

• *The Coastal Cruiser,* by Tony Gibbs, offers the perspective of a sailor who replaced his 37-foot heavy displacement ocean cruiser with a 21-foot racer/cruiser more suitable for the sailing he could plan for the foreseeable future. As editor of *Yachting* for several years, Gibbs reviewed and sailed a wide variety of boats. He draws upon that experience to provide informative commentary on design, rig, accommodations, gear, and safety of cruising sailboats under 30 feet LOA. He includes also his evaluation of eighteen under-30-foot boats.

• *The Proper Yacht,* by Arthur Beiser (2nd edition), is written from the perspective of one whose bias is toward larger boats. The meat of the book is in the first one hundred pages, which Beiser uses to discuss characteristics he believes go together to make up a "proper yacht." In the remaining pages, he offers commentary on 58 boats ranging from 30 to nearly 60 feet in length.

• *Ocean Voyaging,* by David M. Parker, offers the viewpoint of one who strongly favors fast, modern designs of light-to-moderate displacement. Parker does not limit his commentary to boats; he dwells also on the demands and joys of ocean sailing.

• *Blue Water,* by Bob Griffith with Nancy Griffith. Although Bob Griffith died of a heart attack shortly after publication of this book, he left in its pages a legacy of informed opinion molded by some 200,000 miles under sail. Griffith favored the largest boat he and Nancy could handle alone and was a staunch advocate of long-keel designs.

• *After 50,000 Miles,* by Hal Roth, remains today a remarkable assemblage of information and opinion about boats and sailing them. Although Roth does not advocate a particular design approach, he does not hesitate to say what he doesn't like—for example, the IOR influence on many racer/cruiser designs. Moreover, his comments about many of the details required for safe ocean cruising provide important insight into the demands placed on a boat intended for offshore use.

For anyone interested in purchasing a used boat, there is one other book that deserves serious consideration—particularly as you begin to narrow your selection. The book is the *BUC Used Boat Directory* sold by BUC International Corporation, Suite 95, 1881 N.E. 26th Street, Fort Lauderdale, Florida 33305. The BUC book, as it is called, is the best available directory of used boat prices. It is published three times a year and lists the "current" market value of nearly every production boat available on the used boat market in the 25- to 45-foot range. The data are compiled from records of actual boat sales reported to BUC from all around the country. There is also a special edition published about older boats. Although many brokers have found that BUC book price information lags behind the market during periods of rapid inflation, the book is used throughout the industry as a uniform price guide. Specific boat prices, of course, may vary significantly from the BUC book data depending upon the boat's condition, equipment list, and popularity in your local area.

4

Where to Look
Shows, Showrooms, and Docks

Armchair study is a good, low-pressure way to review and learn about boats. Sooner or later, however, you need to move from the living room to the showroom. And to the docks. Hopefully, by the time you're ready to start looking at the boats themselves, you'll have settled on a list of a dozen or so boats you think are worth looking at in the flesh. They are within your size and price range; they appear to offer the accommodations you need; they're of a type you and your mate feel competent to handle; and, their design and construction—so far as you can tell from your study to date—appear to make them suitable for the cruising you want to do. You may even have included one or two boats at each end of the list that appear marginally less or marginally more than you need or can afford. Doing such will help keep your search in perspective and makes shifting gears a little easier if the boats of your choice are not as satisfactory in person as they seem from the photos and rhetoric. You may also have invested a few dollars to consult with a local surveyor about his experience in surveying boats on your list, or in surveying boats constructed by the same builders. The question now is, where can you actually look at these boats?

If you are looking for a used boat, yacht brokers who have listings for boats on your candidate list are the best place to start. If, however, your interest is in a new boat, local dealers who sell boats on your list provide the best starting point. *In both instances, however, leave your checkbook home!* The brokers and dealers are going to try to sell you a boat. But you don't want to buy—at least, not yet. You don't even want to put down a deposit or take a trial sail at this time because doing either one involves some level of personal commitment to a boat that you probably shouldn't be making at this stage of the game. All you really want is opportunity to see a boat that you've read about in the promotional literature and reviews. Moreover, if it has been evaluated in one of the publications you've read, you want opportunity to read through that evaluation while you're sitting on the boat so that you can relate the written words to the actual boat. This is particularly useful if you have the kind of critical review offered by *The Practical Sailor*.

Boat shows also may be a convenient place to survey your candidates firsthand. At the larger sailboat shows, in fact, you may well be able to see all of the boats you are considering at the one show—and all in one day. However, boat shows offer two principal disadvantages:

• The show environment usually makes it difficult to look at a boat for as long and as closely as you'd like. Not only do the crowds of people interfere, but the exhibitors often get nervous if someone examines the boat too closely; that kind of scrutiny is a deviation from the boat show norm.

• Unless you take copious notes, the boats at a show tend to run together in your mind if you look at more than three or four. For this reason, we suggest saving the boat show until you've gone elsewhere to narrow the field to just three or four boats. Then, if the show schedule includes a VIP or special trade day, spend the extra money to go on that special day. Hopefully, in that way, you can avoid some of the crowds.

In contrast, the advantage to starting with yacht brokers or dealers is that most of them will let you poke around and probe their boats to your heart's desire. Above decks, this circumstance lets you take the time to weigh details of design, deck layout, and rigging against your sailing needs. Above and below decks, it lets you weigh the apparent quality of construction against the kind of sailing you'll be doing. It also lets you look closely at the accommodations to study how they might fit your needs. And you don't have to worry about hurting anyone's feelings by the questions you ask or the corners you peer into—

an important consideration when you are a guest looking at the boat of a friend or acquaintance.

Wherever you are looking at boats, however, what you will get out of your survey depends almost completely upon what you bring to it—how well you develop your skills for looking at boats, and the approach you develop for interviewing dealers, brokers, and owners. (A system for looking at boats is described in detail in the next chapter.) Although your approach to interviewing will depend largely upon your personal style, we believe there are two keys to success in this endeavor. The first is self-evident if you put yourself in the position of the person you will be asking to provide you with information. The second stems from the fact that boat salesmen, yacht brokers, and sailors come from a variety of backgrounds and experience. The two keys are:

• *The best approach is straightforward, but a soft sell.* Boat dealers and yacht brokers see all kinds of people and many of them each week. What will impress them is their reading of how serious you are in your intent to purchase a boat. In contrast, they probably will not be especially impressed by your knowledge about boats, or by your aggressiveness in asking questions; some, in fact, may volunteer less and less if you appear to know too much about their product. On the other hand, they are accustomed to prospective customers who don't know much about boats, or about the particular boat(s) they are selling. So you are in a good position to be inquisitive, asking as many questions as you wish so long as the questions are inquiring without being judgmental. The same comment applies when you are talking to sailors you meet along the docks. Sailors are inclined to be helpful because they are generally nice people. However, they are more likely to help people who are genuinely interested in learning than those who appear bent on displaying their own knowledge or who are excessively aggressive in seeking information.

• *You must "qualify" the people you are interviewing.* This boils down to gradually learning about the experience and background each person brings to the table. Some of the most successful salesmen, for example, are exactly that—salesmen; they did not know the pointy end from the blunt end when they started selling boats a couple of weeks, months, or years ago. Others, by contrast, are world sailors who for one reason or another are now on land and need a job. What all of this means, of course, is this: Although some of the boat salesmen, dealers, and yacht brokers may know what they're talking about from personal experience, others have virtually no sailing experience and merely "talk" the language of sailing, tossing around the names of well known sailors featured in the magazines. Or if they do have sailing experience, it may be limited to taking their demonstrator out for a couple of hours for a "demonstration" sail. In such an instance, while the dealer's opinions may be worth listening to if you are looking for a day sailer, they are probably worth very little if you are looking for a coastal or offshore cruiser. In any case, it is up to you to find out what sailing experience the salesman—or the boat owner—actually possesses. Has he been a delivery skipper, for example? If so, what kinds of boats has he delivered, and from where to where? Does he hold a Coast Guard license? Has he crewed on any of the racing circuits? Has he lived on a boat? Or cruised extensively? The answers you receive to such questions will tell you a great deal about the kind of knowledge he brings to his job. It also may suggest whether he has the kind of experience that will be helpful to you. For example, if all of his experience is racing IOR, he may not be the best source of information about a boat for dockside living, or for short-handed ocean cruising.

NEW BOAT DEALERS

Boat dealers—or new boat salesmen (we use the terms interchangeably)—are at the same time some of the best and some of the worst people to talk to about a specific boat. They can be the "best" to talk to because they may have information about the boat that extends well beyond the material you have already received from the builder. These facts often are derived from customer feedback. Dealers can also be the "worst" to talk to, however, because they get paid for selling boats. As a result, they usually are very selective about the content of whatever customer feedback they're willing to pass along. This places the challenge on your shoulders, as the interviewer, to keep the dealer focused on *facts* rather than on opinion. It also challenges you to dig out the unofficial facts by asking gently, but persistently, about customer feedback—including a request for the names, addresses, and telephone numbers of some of those customers.

One underlying fact colors all of any boat salesman's dealings with you, whether that salesman is owner of the franchise or a dealer for the franchise holder: Salesmen usually are paid a percentage of their sales. They may be able to make a "draw" against future sales, but their annual income is a percentage of their sales. In some instances, that percentage is fixed—a commission of 4 percent on the selling price is common. In other instances, the percentage may vary with volume. One large company, for example, has had a commission schedule ranging from 2 to 4 percent, depending upon volume (higher volume yielded a higher commission percentage). What this means, of course, is that every dealer you visit has only one interest—selling you a boat. And you should know what you're up against when you walk into that dealer's office.

In the words of one fairly large Florida dealer, "Most people we see don't know much about boats. That's why we have to have something in our advertisements to attract them. Price is one thing. The rest is just *sizzle*." Boat advertisements, he said flatly, "lie unmercifully about boats. They make claims that bear no relation to the reality of the boat. And we keep changing the message to see what people respond to." The goal of the advertising—or of exhibits in boat shows—is to get you into one of their boats. The next objective, he says, is to get some kind of a cash deposit from you. That deposit, he explains, represents a commitment, even if it is only with a loose contract that provides the buyer with outs. In other words, getting that deposit—even if it is no more than "refundable" earnest money for arranging a demonstration sail—is like hooking a fish: From then on, it's a matter of playing the game skillfully enough to get the fish (you) into the boat.

The actual process of selling a boat to you, this dealer explained, "involves parading all of the plus things in front of the customer while you try to figure out what they think they want. All the time, you are trying to present your boat so that it comes as close as possible to what they want . . . you are trying to find the key to the customer's heart—what is it that makes his eyes shine—so you can appeal to that." The man who makes those statements spent several years as a salesman for one of the most successful U.S. sailboat builders of the 1970s. In the mid '80s, when this interview took place, he and his partners held a major franchise for a large foreign builder. The comments he offered are consistent with our experience.

YACHT BROKERS

Yacht brokers can be an excellent source of information about used boats—or even about new boats if the model has been made for several years and you are trying to assess how well the earlier versions have held up in use. There are several reasons: If they themselves are experienced sailors, brokers often tend to develop specialties that track their own experience and expertise. As a result, it may be possible to find a broker whose sailing experience tracks your sailing interest. Another is that brokers usually represent the owners of a variety of boats rather than a single line. They also have access to boats listed by other brokers through the marine versions of a multilist system. Therefore, they are less concerned than a dealer with which of the boats they sell. This means that brokers are free to discuss the advantages and disadvantages of various boats for different kinds of sailing more candidly than dealers—assuming they are qualified by sailing experience and knowledge of the design and construction of the boats they are showing. (A detailed discussion of the yacht brokerage industry from the buyer's viewpoint and a number of ideas for making the system work for you are contained in Appendix A.)

But the broker is not being helpful out of the kindness of his heart. He, too, has a profit motive. If he feels you are leveling with him and are seriously looking for a boat, he figures he has a good chance of getting the sale in return for his investment of time and expertise in your search. And if (1) you find a broker who will make that investment of his time and expertise in you, (2) you feel good about that broker as a person and develop a good working relationship with him, and (3) you ultimately decide to buy a brokerage boat—even several months down the road—indeed you should buy it through the person who was so helpful to you. In the unlikely event that the boat of your fancy is offered through another office on an exclusive basis, your broker probably can make a special arrangement with the listing broker to go through the boat with you and, if you buy it, he will receive some kind of shared commission for the sale. At the least, you should give him the opportunity to try. It's the only way he has to get paid for his work.

BOAT OWNERS

After you have taken the time to walk around, to go aboard, and to look at and poke around all of the boats on your candidate list, that list probably will be somewhat shorter than it was at the outset. The reason is straightforward: Photographs taken with wide angle lenses, two-dimensional drawings illustrating the interior layout within the maximum length and breadth of the hull, and promotional rhetoric often combine to make a boat look better on paper than in reality. But now that you've measured your list against what you can see by looking at a boat in a showroom or in a slip, it is time to measure the boats on your list against the experience of people who own them. Hopefully, you will talk to several people who own each of the boats on your list. Talking with only one person for each boat will save time, but may be misleading. No individual person's experience with a particular boat is necessarily typical. A bad experience may be the result of poor seamanship or ignorance, not the fault of the boat. Or, that particular boat may be a marine version of a lemon. Similarly, rave reviews may also be the result of ignorance; the boat may be completely unsuited to the use it is receiving.

Cruising World's "Another Opinion" column offers one way to find people who own the boats on your list. Local dealers are another source of names. You can ask dealers for the names and telephone numbers of people to whom they've sold boats like the one you're looking at within the past two or three years. Still another approach involves visiting marinas on any warm weather weekend. When you see someone with one of the boats on your list, just start up a conversation—saying right up front that you are thinking about getting a boat like theirs and wonder if they'd be willing to tell you about their boat.

No matter how you find your boat owners, however, there is one very important step you need to take to make your conversations with them worthwhile. As you did when talking to dealers or brokers, you have to "qualify" the people you're talking to. For example, in the course of your conversations, you need to find out how much and what kind of sailing experience they have. Why? Again, if you are looking for a boat suitable for coastal sailing, what good is an opinion about the boat's suitability for coastwise sailing if the person offering the opinion has never done any coastal sailing? Similarly, if you want a boat for a dockside live-aboard, how much weight should you give someone's opinion about the suitability of his boat for dockside living if he's never spent more than a long weekend on the boat?

However, just because people don't use their boats in the same way that you would doesn't mean you can't learn something about their boats from them. Far from it. Knowing how they use their boats can guide both your questioning and your evaluation of responses you get. For example, if a person spends most of his time motoring and very little sailing, it may be worthwhile exploring the make and size of engine, its maintenance and repair history, its suitability for the boat (cruising speed in knots and the cruising speed RPM), the propeller size, and anything else you can learn about the installation and the owner's experience and satisfaction with it. Or, how does the boat handle under power in tight maneuvering? And, what techniques has the skipper developed for maneuvering in reverse?

You can ask similar questions about their experience with their stoves, compasses, winches (are they standard size? If not, why not?)—even about the placement of cleats and chocks. It's a chance to pick brains and you should use it. Most sailors are friendly. They're not only willing to talk about their boats, but the more questions you ask, the more they enjoy it. In fact, about the only thing the sailors we know seem to enjoy more than talking about their boats and their sailing experiences is actually being out on the water sailing. Moreover, many sailors will recognize themselves in you. They've stood in your shoes too—trying to choose their own boat.

Of course, the question in your mind must be: "Will the guy who owns one tell the truth about his boat?" The answer is, "Yes, but. . . ." The qualification is this: Sailors generally do not know their own limitations. Almost always, they believe in their own boats, or they wouldn't be sailing them. In one word, sailors are "optimists." And there are two things you must do to screen out the effects of their inherent optimism: (1) Talk to several owners of the same kind of boat so that you can form a composite picture based on the experience of a number of people—each hopefully with different kinds of sailing experience—rather than relying solely upon the experience of just one person. (2) Explore each sailor's experience with his boat and its systems, rather than seeking his opinion. For example, it is one thing to know that someone thinks his boat is a good heavy-air sailing machine; it's another to know that the strongest winds he's had it out in gusted to 25 knots. Or that he only sails downwind in heavy air. Alternatively, if he offers the opinion that his boat sails well in light air but tells you that he starts up the engine every time the wind drops below eight knots, you can question whether he actually knows whether the boat sails well in light air. He may just be repeating what the salesman told him. He may also be correct, but that will come out in talking to other owners if, in fact, it is true.

The more specific your questions are, the more you can learn. How many days does he sail every month? Where does he go when he takes the boat out? What kind of cruises does he take? Does the spouse sail with him (her)? If "Yes," under what kinds of conditions? How does the boat balance under sail? (Very important if you hope to use a wind vane.) How would he change his boat to make it better? Or, if he inherited a large amount of money, would he get a different boat? If so, what would he look for that his present boat doesn't have? How much longer does he expect to keep this boat? Where did he buy it? Did the boat need any work to put it into shape before delivery to him?

Those are but a few questions, but the pattern and purpose should be obvious. You need to learn both about the sailor and the boat. In this way, you can judge the validity of his comments. One added note: As valuable as conversations with sailors about their boats can be, conversations at the yacht club bar are invariably a waste. Anyone on a bar stool can be the world's most experienced sailor. Truth is found much closer to the dock. And on that sailor's boat.

5

How to Look
A 15-Minute Checkup

In all likelihood, as important as the information is that you glean from interviewing other people, what you learn from your own observations will have the most impact on your decision-making process. For that reason, the technique to be developed in this chapter for evaluating a boat is among the most important skills you can learn in the process of choosing your boat. It requires only fifteen minutes or so to develop a surprisingly detailed analysis of the kind of sailing the boat is best suited for. Moreover, this same system can be used whether you are in the midst of a crowded boat show, are having a leisurely look into every nook and cranny of a boat in a dealer's showroom, or are looking discreetly at a friend's boat while you are his guest for a Saturday afternoon sail. In any case, this evaluation will not only enable you to prune your list of candidate boats, it will also serve well as the basis for the more detailed look you will want to take later on at the boats that survive this overall screening process—the semifinalists, so to speak, in your selection process.

This approach to reviewing boats was developed initially in the 1970s for quality-control inspection purposes at the former Westsail plant in North Carolina. Since that time, it has been expanded to make it applicable to virtually any fiber glass sailboat. It has also been expanded to look at function—the kind of sailing the boat is suited for—as well as quality. Both are areas of critical importance to you in choosing your boat. This is not a technique, however, that you can read about and then simply put to work. You must practice it before looking at the boats on your list. You can—and should—practice it on every boat you see. This practice not only lets you develop your powers of observation; it also helps you to build a set of reference points that you can use to judge what you see—for example, the relative size of a boat's rig and rigging, or the adequacy of hardware.

It's also worth noting that there's an added bonus from developing skills in this area: That is, the techniques you develop for evaluating the strengths and limitations of different boats can bring an entirely new dimension to your pleasure in boat watching—a dimension that will add to your enjoyment of boating for the rest of your life. You may even discover, as we have, that a local marina is the first place you turn for entertainment when visiting a new city. We often find, for example, that walking along a dock looking at boats and, if we're lucky, finding other sailors to talk with provides the best show in town. It can be the same story for you, too: The more you know about the boats you're seeing and, as discussed in Chapter 4, about interpreting what their owners tell you about them, the more enjoyable walking the docks becomes.

No 15-minute walk-through, of course, will yield a detailed marine survey. Nor should anyone expect to produce in 15 minutes the kind of detailed information about the boat's design and construction that you could develop if you took several hours to look at each boat. Fifteen minutes, however, is ample time to form a reasonably accurate impression of the boat. And that, after all, is all you are trying to do at this point: You are trying to form a solid enough impression of each boat to whittle your list of a dozen or so down to two or three—possibly four—that you believe merit a second, more thorough look. So in a single word, you are dealing with *impressions:* How well suited does this boat's design and construction seem for the use you would make of it? What appears to be the quality of workmanship? How much attention has the builder paid to detail? It is the answers to these questions that will help you determine whether to keep each boat on your list, or to drop it.

In one sense, the system we are suggesting is analogous to

working a jigsaw puzzle. Each piece of the puzzle contains a detail of the overall picture. The more of these pieces you fit into place, the more complete the puzzle's picture becomes. Each boat you look at becomes a new puzzle for you to solve. Some of the puzzle pieces show details of design features that suggest what use the naval architect had in mind for the boat when its plans were drawn. Others involve details of construction or equipment that suggest the use the builder expects the boat to receive (not necessarily the same use for which it is advertised). Still other details speak to the quality of thought and workmanship that have gone into the boat's construction. The basic notion is that a boat can be designed for one use (e.g., dockside living), apparently constructed for an entirely different use (e.g., crossing oceans), and then be of such poor quality of workmanship to make it a questionable choice for either purpose. In any case, the technique developed here involves looking at details. It involves taking visual snapshots of design and construction details, making notes in a small notebook about those snapshots, and then fitting those snapshots together like a jigsaw puzzle to form a mosaic which represents your evaluation of each boat's suitability for the kind of sailing that you want to do. Keeping the notebook is a key to the process; without the notebook, you'll forget the details you observed on one boat the minute you start looking at the next boat.

THE EXTERIOR

As in everything you've done so far in this process of choosing your boat, the most important element in evaluating boats lies in being systematic. That's why we suggest you begin looking at each boat from the outside. Not only can you be looking for exterior details of interest, you can also be noticing elements on the outside that you want to check from within.

The first step involves backing away from the boat and looking at her lines and rig. *Is she pretty?* If you don't think so, you might as well move to the next boat. Few people will be happy with a boat they do not think is pretty.

Does the rig appear proportional to the rest of the boat? If not, you may want to look closely at the sail area to displacement ratio. If the rig looks undersized to your eye, it probably is undersized—not a problem if your objective is dockside living, but a possible source of frustration if you plan to do much sailing.

Moving closer, *does the standing rigging look sturdy? Or, does it seem on the light side?* Racers often want the lightest standing rigging they can get away with to reduce both weight aloft and windage. Cruising boats, on the other hand, are often better served by a larger safety margin in rigging size. In general, however, rigging should be sized to meet the demands that will be placed on the boat, and those demands will depend upon the kind of sailing you are doing. For ocean sailing, we like to see sturdy, stainless-steel wire rope. We also prefer the stainless wire over rod rigging for offshore sailing because of the relative ease of repairing the wire rope system. For coastal and inshore cruising, where rigging failure is less critical, lighter wire may be satisfactory. Rod rigging also is suitable for coastal and inshore waters because repairability is less crucial to safety. For day sailing, where help is always nearby, the wire can be lighter still, though it may prove expensive if the rigging fails in a sudden squall.

How are the stays and shrouds anchored? Do chain plates for shrouds go through the deck or bolt to the hull? (You'll want to check below for evidence of leakage around chain plates if they go through the deck.) Does the headstay fitting fasten only to the deck (unsuitable for any but protected waters)? Or is it tied securely into the hull? How is the backstay fastened at the stern— to the deck, hull, or both? For any but sailing in protected waters, the backstay too should be attached to the hull, or to the hull and deck (as with a chain plate knee).

Does the boat have single or double lifelines? Are stanchions strongly anchored? (Photo 1.) Single lifelines are O.K. for dockside living or sailing in protected waters, marginal for sailing in semi-protected waters, and flatly unacceptable for coastal or offshore sailing. Double lifelines should be demanded for coastal or offshore sailing. Similarly, bow and stern rails on boats intended for coastal or offshore use should have double rails.

If the boat has high bulwarks, are there large drains for water to escape the deck? High bulwarks provide a sense of security, but if

1. Lifelines are only as effective as the stanchion base is strong. The two-bolt attachment shown is probably adequate for protected or, in good weather, semiprotected waters.

they can trap a large volume of water on deck in heavy weather, the boat's stability may be compromised. For that reason, such bulwarks may be questionable for coastal sailing and, possibly, unsuitable for offshore work.

Are windows large or small? If they are large, is there provision for installing storm shutters to protect them in heavy weather? Also, how are they installed? In general, large windows in sailboats of the 25- to 45-foot range are not suited for coastal or offshore sailing unless they are fitted with sturdy, wood shutters to keep them from being broken in heavy weather. Conversely, small fixed ports are unsuitable for dockside living. In addition, bolted windows are preferred for coastal or offshore work.

Are opening ports made of metal or plastic? If plastic, the boat probably is not suited for use beyond protected or semiprotected waters.

On the foredeck, *is there an anchor well?* If so, *is it adequately drained?* We have seen boats at boat shows with anchor wells

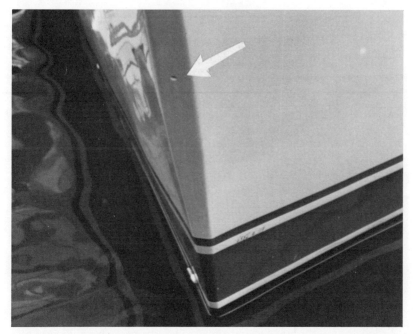

2. The anchor well on this boat drains through the pencil-size hole in the bow.

half-full of water for one of two reasons. Either the builder forgot to install the drain, or the drain was so small that it was easily plugged by dirt or debris. Such situations make one wonder about the quality of thought that went into producing those boats (Photo 2).

Is there an anchor roller? If so, is it made of sturdy stuff? Or is it made of thin sheet metal or lightweight castings—materials unlikely to withstand high stresses generated as the boat pitches when anchored during a storm? Is there a hefty pin to hold the anchor rode or chain in place under storm conditions? (Photo 3.) Any number of boats have broken loose in foul weather when their anchor rollers failed under stress, or when the rode jumped out of the roller and was chafed to breaking. Under the assumption that any sailor may someday find himself at anchor in a poorly protected harbor when a thunderstorm or squall line comes through the area, our own bias favors sturdy

3. An anchor rode will quickly jump off this bow roller and begin to chafe at the base of the bow pulpit brace. Alternatively, the rode must be led around the bow rail and run through the deck chock. We would also be concerned about the vulnerability of a bow rail which extends so far in front of the boat.

anchoring gear for any boat. However, one could argue that lightweight anchor rollers are adequate for a boat intended only for dockside living or day sailing in protected waters.

Are there sturdy handrails you can hold onto when moving about on deck? The old dictum "One hand for yourself, one for the boat" is meaningless unless you have something sturdy to hold onto when moving about on deck. And a lifeline does not qualify. Any boat used for all but the calmest sailing should have sturdy handrails inboard to hold onto when moving around on the deck.

When you walk on deck, does it feel solid? You can hear voids in the laminate as creaking or crackling sounds.

Is the cockpit large, medium, or small? In general, large cockpits are intended for socializing at the dock or when anchored in a protected river or cove. They are not intended for offshore cruising. Small cockpits are intended for offshore or coastal sailing, and are poorly suited for entertaining. Moderately sized

4. This drain would be hard-pressed to keep up with the full flow of water from a garden hose.

cockpits are an obvious compromise. Similarly, you should check the cockpit drains. They should be capable of draining the cockpit quickly if the boat is intended for coastal or offshore use (Photo 4).

Are there sharp corners on the cabin house or in the cockpit that the crew can fall against if the boat lurches suddenly? The answer should be, "No." If there are many sharp corners, the boat's use in any but protected waters should be questioned.

Do any cockpit lockers open directly into the cabin area? If so, the design is probably intended for use in protected or semi-protected waters. Such lockers can let large amounts of water into the boat if the cockpit is ever flooded.

Does the companionway opening extend all or most of the way down to the cockpit sole? Alternatively, is there a bridge deck to the height of the cockpit seats? A deep companionway opening suggests the boat is designed for use in protected or semi-protected waters (Photo 5). The bridge deck is important for

5. The low bridge deck makes it easy for crew and water alike to go from the cockpit into the cabin.

coastal or offshore sailing to help keep water from getting into the boat from the cockpit in heavy weather.

Is the companionway hatch protected by a sea hood? If not, the boat pretty clearly is not intended for coastal or offshore use, and probably not for use in semiprotected waters.

Are running lights of a good size? Or are they the minimum necessary? Also, are they sealed where the wire goes into the light fixtures? Unsealed lights will fail quickly from corrosion. Whether a builder seals running lights against the water or leaves them unsealed provides commentary on the attention to detail (quality) that goes into the boat.

Are fuel and water tank fill plates next to each other? Or, are they located on opposite sides of the boat? Are they clearly labeled? More than one boat owner has pumped fuel into his water tank or water into his fuel tank—in part, at least, because the fill pipes were unlabeled or were located so that it was easy to make the careless error.

THE INTERIOR

As you move down below, you need to do three things: (1) Keep in mind the items you noted on deck that you want to check from below; (2) continue looking for design, construction and quality factors to help classify the boat; and (3) pay particular attention to the use of space—how much there is, and how it is used.

It sometimes helps to start at the bow and work aft. Again, it is a matter of being systematic. As you move forward, however, imagine that the boat is heeling over some 20 to 25 degrees, and that it lurches suddenly. Then go through the same exercise again, but with the boat heeling in the other direction. *Is there something to hold onto at all times? How far could you fall? And, what would you fall into?* Large cabins are delightful for living aboard at the dock or for use in protected and, possibly, semiprotected waters. For coastal cruising and offshore work, however, you want plenty of good handholds, and short distances to fall. The farther there is to fall, the more risk of injury. Also, sharp corners or edges have no more place down below than they do above deck—at least, not for any but dockside living or sailing in protected—possibly semiprotected—waters.

As you move forward, you also want to look at the general layout and use of space to get an overall impression of whether the interior seems as suited in reality for the use you want to make of the boat as it sounded in the promotional brochures. Your response to this question at this point is important: If you are struck in passing through the interior that the boat's layout simply will not work for you, you may want to remove the boat from your list. But don't stop your evaluation; you need the practice.

Once you reach the area of the chain locker, you are ready to start looking at details. For example, if you look up into the forepeak (this is where a small flashlight often is helpful) you should be able to look at the hull-to-deck joint. Even on boats built with hull liners and / or anchor wells that open to the deck, the joint should be accessible in the forepeak. There you can start finding the answers to your questions:

What kind of sealant or adhesive sealant was used in the hull-to-deck joint? We have seen one $80,000-plus boat in which a puttylike substance was used. In a relatively short time, that putty will harden, crack, and leak. The sealant should be tough, but rubbery; 3M's 5200 adhesive / sealant is excellent.

How is the joint fastened mechanically? With self-tapping screws, bolts, or a combination of screws and bolts? If bolts, are they fastened with flat washers and either aircraft nuts or nuts and lock washers? Also, how closely are the fasteners spaced? These questions may seem unimportant, but we have seen many boats at boat shows and in marinas in which there were neither aircraft nuts nor lock washers in the hull-to-deck joint. We have also seen a substantial number of boats in which workmen failed to put any nuts on the bolt ends in the forepeak—probably because it was difficult to reach up there. Additionally, we have seen bolt ends so close to the side of the hull that there wasn't enough room to put on the nuts. When there is that kind of lapse in quality control on such a critical part of the boat, one should be concerned about possible lapses in quality that are hidden from view. Sometimes, too, there may be too many fasteners, weakening the deck or hull flange in the same way that a line of perforations makes it easy to tear a piece of paper. In general, bolts should be spaced every four to six inches.

How is the headstay fitting you looked at from the outside fastened on the inside? It should be bolted, with backing plates and lock washers or aircraft nuts.

Is the deck core sealed around the chain pipe? Wherever there are holes in a deck—even for bolts used to attached fittings—the sides of the hole should be sealed to prevent water from entering the laminate or any wood core used in the deck. The hole cut to run the anchor rode into the chain locker provides one of the few places you have to check whether the builder thinks about sealing the laminate. Even if a small chain pipe is fitted, it is often possible to see whether the edge of the cutout in the deck has been sealed by looking up from the chain locker. If it has not been sealed either with gel coat or, preferably a sealant such as 3M's 5200, there is again a suggestion of poor attention to critical detail. On the other hand, the builder whose workmen do seal the laminate properly are demonstrating a positive approach to quality.

How are such foredeck fittings as bow cleats, windlass, and bow pulpit fastened to the deck? They should be through-bolted with sturdy aluminum backing plates. An alternative for the windlass may be a large, but thick (¼-inch or more) fiber glass backing plate molded to fit the curve of the deck. These bolts ends, too, should be fitted with either aircraft nuts or lock washers and nuts. The metal backing plate serves as its own "flat washer." If a fiber glass backing plate is used, a large flat washer is needed beneath the lock washer or lock nut.

Is there an anchor well that opens to the deck? If so, it is doubly critical to check the hull-to-deck joint and method of fastening deck fittings. If the anchor well is molded as part of the deck molding, there must be adequate room on the sides and front for someone to install fasteners both for the hull-to-deck joint and for any fittings (including the stem head fitting) that cannot be installed before the deck is put on the boat. In addition, there should be two hose clamps (double hose clamps) at each end of any hose or flexible tubing used to drain the anchor well.

Is there an eyebolt or ring for securing the bitter end of the anchor rode? More than one sailor has watched in dismay as the end of his anchor rode went overboard while he was letting out more line in a squall and didn't realize that he was so near the end of the rode. Surprisingly few boats are equipped with such a fitting; as a result, when a boat is so equipped, it provides positive comment about the builder's attention to detail (Photo 6).

6. A too seldom seen feature the U-bolt in the anchor well or chain locker for securing the bitter end of the anchor rode.

How is the bulkhead dividing the forepeak and the forward cabin attached? Often, because of the tendency of builders today to cover up everything with veneers and fabrics for decoration and cost-savings, the forepeak side of this bulkhead is one of the few places on the boat where you can see how bulkheads are installed. (Two other places in modern boats where you can often see bulkhead attachments are from inside the lazarette and/or large cockpit seat lockers.) For that reason, it is important to look here for this detail. If the boat is built using a full hull liner, this bulkhead may be a molded part of that liner. It may also be either constructed from plywood or a flat composite sheet fit into a flange or groove in the liner. In the latter instance, it should be fastened in place mechanically—for example, with self-tapping screws—so that it can't pop out as the boat works in heavy seas. If the bulkhead is integral to the liner, it should have molded-in ridges or other stiffeners so that it does not flex in heavy seas. Otherwise, the boat is best limited to protected or semiprotected waters. If the bulkhead is bonded directly to the hull using fiber glass materials, there should be a fillet (wedge) of foam between the edge of the bulkhead and the hull. The fiber glass taping should extend at least six inches onto the wood bulkhead and four inches onto the hull. Lesser bonds suggest protected or semiprotected waters. Finally, in all instances—though it is seldom done—the bulkhead should be bonded to the deck.

If the boat has a vee berth forward, *is it usable as a berth for two adults?* Too many boat builders have provided vee berths that are too short and come to such a point at the forward end that there is no possible way for two normal-size adults to sleep on that berth at the same time with any degree of comfort. As a result, vee berths have gotten a bad name among some sailors. However, vee berths can be both practical and comfortable. All that's needed is a berth that is long enough (approaching seven feet) and wide enough (a foot or more wide at the pointy end).

How is space beneath the forward berth used? This question involves stowage and weight distribution. The areas beneath all berths in a boat offer large potential stowage space. Often, however, this space is taken up with water tanks or drawers. In a relatively light displacement boat, a water tank under the vee berth may put more weight forward than is desirable. Water weighs about 7.5 pounds per gallon, with the result that a forty-gallon water tank represents about three hundred pounds of added weight—possibly enough to affect the boat's balance and stability. In a boat of heavier displacement, such a tank may not be a problem. In either case, however, the tank robs stowage space—something of interest whether you plan to live on the boat at dockside or use it for serious coastal or offshore cruising.

Do shelves have sturdy, high fiddles to keep things on them? This question is important throughout the boat for any but dockside living. However, the need for restraints of some kind to keep items on shelves is greatest in the forward cabin, where the boat's motion is most exaggerated. In addition to being high enough, the fiddles also must be sturdy enough that they won't break if someone grabs them when the boat lurches.

Do cabinets have shelves or drawers? If stowage space is particularly important, shelves are preferred. Drawers—though convenient—waste a significant amount of space. They also are more expensive if they are made well. Shelves, on the other hand, even in cabinets, need good, high fiddles to keep contents from falling out when the door is open. A strong, positive latch also is needed on cabinet doors.

As you feel around the inside of cabinets, do you find any sharp edges? The insides of cabinets and lockers are often "out of sight" and "out of mind" as the boat is being built, with the result that workers may forget things. Consequently, these are excellent places to look for either good attention to detail (smoothly finished surface) or poor attention to detail and the subsequent rough finish. Similarly, where hinges or latches are attached, look for the sharp points of screw ends. Clean, smooth surfaces tend to indicate good workmanship. Rough or sharp surfaces, and screws that come through the wood to snag either your fingers or whatever is stowed in the locker, suggest mediocre workmanship.

If the boat has hanging lockers, *are the lockers deep enough and tall enough to hang up the usual garments that need hanging?* At the same time, if you will be sailing coastwise or offshore, is there some way to tie the hanging garments together against the side of the locker to prevent them from swinging back and forth, constantly chafing?

Do cabinets and lockers have adequate ventilation? It is difficult to define "adequate" in the abstract. However, as a rule, all cabinets and lockers that contain clothes, sheets, blankets, towels, etc., should have sufficient ventilation through the sides, doors, or drawer fronts to prevent mildew.

Moving aft into the head area, *do all hose fittings have two hose clamps at each connection point?* Usually, you can add hose clamps, but the builder's use of hose clamps says something about his concept of quality. *Do through-hull fittings have sea cocks or gate valves? Are those sea cocks (or gate valves) easily accessible?* And, *are the through-hull fittings bonded to a common ground?* Many sailors—including the two of us—consider sea cocks mandatory on boats for use in semiprotected, coastal, or offshore waters. The use of sea cocks on through-hull fittings is less critical for day sailing and dockside living, but the vulnerability of gate valves to corrosion argues against their use on any boat—even those that will never leave the dock. As for accessibility, if you can't reach a sea cock (or gate valve), you can't shut it off in an emergency. We prefer sea cocks so easily accessible that they can be turned off conveniently whenever you leave the boat. In that way, a failed hose or clamp cannot sink the boat in your absence. In any case, all through-hull fittings should be connected by a length of copper wire to a common ground with the engine to help prevent degradation of the through-hull from electrolysis.

While in the head, note also where the sink is located? *How far above the waterline is the sink? And how far outboard of the boat's centerline?* As noted earlier, if a sink is only marginally above the waterline, or is located outboard so that it can be put below the waterline when the boat heels, you need to be concerned about possible flooding as a result of water siphoning back through the sink into the boat. For use offshore or in coastal waters, the sea cock on the drain for such a sink should be kept closed except when the sink is being used. *If the toilet has a connection so that it can be discharged directly overboard, what is the toilet bowl's position relative to the waterline?* To prevent back-siphoning through the toilet, the head discharge line should be looped well above the waterline (right up to the sheer) and an antisiphon valve installed. If the top of the bowl could be below water level if the boat is heeled well over (and that's most boats), the shut-off valve for the discharge line should be readily accessible so that it can be kept closed except when the toilet is in use to protect against flooding if the antisiphon valve fails.

How is fresh water pumped to the sink? Hand or foot pump? Pressure water system? Both? (The same questions apply to the galley sink.) For dockside living and relatively short distance cruising, pressure water systems are delightful. For coastal or offshore sailing, there should be a manual pump—preferably, a foot pump—system for delivering fresh water. The manual pump serves two functions: (1) It serves as backup to the pressure water system when it fails (as it will); and (2) it makes it easier to conserve water. Pressure water systems encourage waste—not a problem if you are either connected to a shoreside water supply or able to fill your tanks as needed; using more than the minimum amount of water required at any given time is a potentially serious problem, however, if you are crossing an ocean or making a coastal run of several days and your water supply is limited to the contents of your water tanks.

It also is now worth beginning to look at wiring—perhaps in the back of cabinets. *Is it neat?* In general, wiring should be supported every six to eight inches, but run loosely enough to allow for hull or deck movement without stressing the wires. *Does wiring run behind a hull liner?* Or, *does it run between a fiber glass headliner and the deck overhead?* Preferably, all wiring will be run so that it can be repaired or replaced easily if necessary. All too often, wiring is run before hull liners or headliners are installed and is completely inaccessible for repair or replacement in the finished boat. Also, wiring between the hull and a hull liner may be vulnerable to damage from chafe. Malcolm and Carol McConnell tell in their book, *First Crossing: The Personal Log of a Transatlantic Adventure,* of fire starting in the hull liner of their boat (fortunately, not the hull) halfway between the east coast of the United States and the Azores when wires between the hull and hull liner shorted out from being chafed as the boat worked in gales offshore.[1]

1. " 'Mal!' she cried again. 'It's . . . there's something burning down here.'

"When my boots struck the cabin sole, I could smell the unmistakable odor of burning epoxy resin, a sour, chemistry-class stench that had the hair on the back of my neck literally standing erect. I snatched the fire extinguisher off its bracket above the chart table and dragged open the accordion door. Carol was sitting upright on her bunk, her

What provision is there for ventilation in bad weather? Ventilation is needed whether a boat will be used for dockside living or ocean cruising. Virtually all of us will be aboard our boats more than once when it is raining and all hatches and opening ports must be closed to keep dry. So there should be provision for ventilation even when hatches and ports are closed (Photo 7).

7. Although dorade vents provide a flow of air below deck in fair weather or foul, they are notorious for snagging jib sheets. This builder has gone the extra step of providing a stainless guard wire to keep sheets clear of the vent.

legs still in her sleeping bag as she pulled at the zipper. The choking smoke was now visible in the cabin, illuminated in broken layers by the chart-table light. Her arm shot out, pointing directly at the opposite side of the cabin, at the white fiber glass liner just below the forward end of the tear-shaped window.

"I stood fascinated by the malignant oval of molten gelcoat. In the half light, the brick-red glow was plainly visible. The thick bridle of lighting wires was under the liner at that level. Obviously there'd been a bad short circuit. . . .

"I now understood what had caused the short and the fire. The severe pounding the boat had taken when we'd lain ahull in the first gale then sailed under storm sails in the next two blows had *torqued* the boat, putting the hull under enormous strain. The plastic hull being flexible by design, parts of it moved. Undoubtedly there'd been some slippage between the hull and the liner, and this had chafed the wires, eventually producing the short." Malcom and Carol McConnell, *First Crossing: The Personal Log of a Transatlantic Adventure* (New York: W. W. Norton & Company, 1983), pp. 139–40 and 144.

Is the mast deck-stepped or on the keel? If stepped on deck, there should be a sturdy compression post between the underside of the cabin top *and the keel* to support the mast step against the downward pressure created by the tight stays and shrouds. If the mast is stepped on the keel, is it really stepped on the keel, or is the mast step a bit forward of the keel? In some modern fin-keel designs, the mast is stepped on the hull itself, not on the keel. Particularly if the rig is tightened down hard, these boats may develop leaks in the hull around the mast step because the hull is asked to carry a load it was not constructed to handle.

What size plywood is used for structural bulkheads and, where applicable, for furniture? In boats of about 30 feet and longer, all structural bulkheads should be either three-quarter-inch plywood or an equivalent-strength composite material. Most furniture elements, if made of wood, should be constructed using half-inch plywood, or the equivalent composite sheet structure. In smaller boats, half-inch plywood or the equivalent composite material should be used for structural bulkheads. For dockside living, protected and semiprotected waters, a lighter weight plywood can be used for furniture elements, particularly in the smaller boats. However, under no circumstances should structural bulkheads be less than suggested and, in 40- to 45-foot boats, double thicknesses of plywood (1½ inches thick) may be preferred for either offshore or coastal use.

As you continue to work your way aft, you will be looking for many of the same details noted in the forepeak, forward cabin, and head. Lockers all should be ventilated in some manner. And, as you look into lockers, you should continue to check routinely (and carefully!) behind hinges and latches to be certain the sharp ends of screws are not protruding. You should also try again to look at the hull-to-deck joint. You may be able to look again at the installation of bulkheads—how they are attached. Hatches and windows should be checked for signs of leakage. Accessibility of shut-off valves for through-hull fittings also should be checked.

Where winches and fittings, including handrails, are fastened on deck, are the undersides (the bolt ends and backing plates) accessible from inside the cabin? Or, are they covered essentially permanently by the headliner? Do all deck fittings have backing plates? If not, the builder apparently is cutting corners. He is also cutting corners if the undersides of deck fittings are not accessible and you can't determine whether backing plates were used. Apart from safety, deck fittings may develop leaks and need to be rebedded—somewhat difficult if you can't remove them short of major surgery to the headliner.

Do shelves and counter tops have adequate fiddles to keep objects from sliding off when the boat heels or pitches? Fiddles should be a minimum of one inch—higher on boats for semiprotected, coastal, or offshore use—and their corners rounded so that dirt can't get stuck in a sharp corner. As above deck, any corner should be rounded so that it is less hazardous if you should fall against it.

Where are the main tanks for water and fuel? Do they have clean-out plates? *Are there readily accessible valves on top of the tanks to shut off the water or fuel?* If the water tanks are beneath settees, they may be taking up important stowage space for living aboard or long-distance cruising. If the fuel tank is beneath the lazarette or in some such out-of-the-way place, you need to look carefully at accessibility for maintenance of fuel lines and, possibly, for cleanout. In any case, failure to provide clean-out plates or shut-off valves on the exit lines in fuel and water tanks represents an unacceptable means of cutting corners just to save a few dollars.

Looking at the galley, *how close to the centerline is the galley sink? How large and deep?* The galley sink potentially provides the same risk of flooding the boat as the head sink if it is too far off the centerline and/or near the waterline of the boat. For that reason, it should be close to the centerline for any serious sailing. Size and depth of the sink (or sinks) also is important; if it is large enough and deep enough to hold your biggest pots, the sink provides a safe place to put hot foods when cooking while underway. If the sink is shallow, it's really only intended for use in protected or semiprotected waters.

For added safety, there should also be a hefty bar or railing in front of the stove to keep the cook from falling against the stove. *Is the icebox adequately insulated?* The icebox should have a generous four inches of foam insulation on all sides plus top and bottom—and that includes the opening top as well. Unfortunately, too many builders—either to cut costs or to provide you with the largest possible icebox—put no more than half of the recommended amount of insulation around their boats' iceboxes. *Does the icebox drain into the bilge?* Hopefully not. The icebox drain should be run either to a special closed sump that can be emptied periodically, or to a pump for the melted ice to be pumped directly from the icebox.

As you look at the engine, *is it easily accessible for service underway?* Again, the importance of accessibility increases with the level of demand placed by the type of cruising you'll be doing. Accessibility is most important when you need to change or add oil, change the water-pump impeller, or replace an alternator drive belt while the boat is at sea. It is considerably less important when all maintenance and repair will be done at a dock.

Two final notes:

• After reading this list of details to check, one can reasonably ask how it's possible to look at all of these details and make notes on them in 15 minutes. The answer clearly is . . . it's not possible. However, in 15 minutes you can look at and note enough of these kinds of details to gain a pretty good sense of the boat, its quality, and the kind of use it is suited for. Moreover, as you gain experience, you'll find yourself tending to focus on a consistent set of details as you go through each boat. In addition, you can help yourself by developing a system for taking notes. One such note-taking system involves making a checklist of all of these items and leaving room opposite each for a few notes. Another involves simply making three columns across the page—Pluses, Neutral, and Minuses—with plenty of room for elaboration. As details are noted, a value judgment is made as to which column is appropriate and details are noted in that column. In any case, the key is to take notes.

• Although the system we have outlined for looking at boats is a good one, we would be the last to claim that our way is the only way. Or that it's necessarily the "best way." It is, however, a good system developed to be useful under a variety of circumstances. It also has enough built-in flexibility that its elements can be adapted to your own personal needs and style. Most important, it is a method for evaluating boats that can be learned easily by anyone willing to make the effort to do so.

II

Fiber-Reinforced Plastic (Fiber Glass) Boat Construction and Design

It is one thing to use a screening technique to narrow your selection to a chosen few. It is quite another to take the next step of making a reasonably thorough—even if not expert—evaluation of the remaining candidates to narrow the choice to one or two finalists. One way to do this is to hire the services of a marine surveyor and have him (or her) evaluate each boat for you. Another is to prepare yourself to go over each boat thoroughly by learning what you can about boat construction and design, reserving the possible expense of the surveyor as a last check on the boat you've tentatively chosen.

6

Fiber Glass Boat Construction
A Look beneath the Gloss

On a perfect day, you can probably take any sailboat out of inlet "A" into the ocean and bring it back some time later through inlet "B." Unfortunately, many people do just that, and then talk about it. Considerably more often than not, however, the boat they used that day was not well suited for coastal or offshore sailing. And they were lucky; the weather held as expected. The Fastnet race in 1979, the Farallones race in 1982, and the Fort Lauderdale to Key West race in 1983, however, all were well-publicized examples of occasions when the weather did the *unexpected*. Those races also became examples of people being in the wrong place at the wrong time with the wrong boat. Lives were lost in the Fastnet and Farallones races. Boats were lost in all three.

If these examples were rare occurrences, we could perhaps put them down to the unusual. Unfortunately, more such stories are appearing every year in the newspapers and sailing magazines. Moreover, with the possible exception of the Fastnet race, the conditions encountered in those races are not all that unusual in ocean waters, and boats intended for ocean cruising—including coastal cruising—must be designed and constructed to take care of their crews in such conditions. Even then, it may not be quite enough. The only Westsail we know of that has been sunk in the open ocean is a Westsail 32 lost while returning from Bermuda in early summer weather patterns. A severe storm arose quickly one night, building up very steep waves. Before the owner could reduce sail, the boat went through three uncontrolled jibes and a severe knockdown—the combination of which cracked the hull on the starboard side just above the waterline. The typical Westsail 32 hull has six layers of 1.5-oz. mat and five layers of 24-oz. woven roving. The laminate is about one-half inch thick. At the same time the hull was being cracked, the bulkheads broke loose from the hull. They had been bonded using two laminates of 1.5-oz, mat and 7.5-oz. boat cloth extending four inches onto both the wood bulkhead and the hull.

At this point, the owner was able to keep up with the water leaking through the crack in his hull by using his bilge pumps. When the weather eased, he thought he would be able to limp on to his home in New Jersey despite the fact that his rigging was slack because of the loose bulkheads and could not be tightened without opening up the crack in the hull. After three more days, a second storm arose that opened the crack in his boat's hull even more, until water was coming in too fast for him to pump and he turned on his EPIRB. He, his wife, and young daughter were rescued several hours later by a ship directed to him by the Coast Guard. His boat, constructed to sail the world, sank beneath the waves.

When the first fiber glass (FRP) sailboats suitable for cruising were introduced in the mid-1950s, they generally were built so heavily that some of their owners today refer to them fondly as "icebreakers." The hulls were typically constructed to wood scantlings. No one knew just how good those newfangled glass-fiber and plastic-resin building materials were. Nor were boat designers or builders much accustomed to "engineering" boats—developing laminate schedules and hull reinforcing systems to optimize the balance between panel strength, panel stiffness, and weight. Quite the opposite, in fact. The construction of wood boats had been refined for centuries, developing a large amount of conventional wisdom to guide both designers and builders in the details of specification and assembly. As a result, the focus was principally on boat *design*—beauty, sailing ability, and, perhaps, accommodations—rather than on materials engineering.

But that was yesterday, a "generation" ago in the fullest sense of the word. Today, the continued development of FRP technology, the escalation in materials and labor costs, the growth of a world-wide boating industry to supply the U.S. market, and the intense competition for our boating dollar have combined to make materials engineering an integral part of boat design and construction—particularly in the United States, Europe, Japan, New Zealand, and Australia. The result is a broad spectrum of construction practices. At one extreme are the designers and builders who are engineering boats to the minimum "acceptable" parameters; at the other are designers and builders who are still producing modern-day versions of icebreakers. All have their place; the challenge is for you to develop enough knowledge to know where on that spectrum you want to find your boat. More than any other single factor, the construction of the boat you buy will describe the reasonable limits of your cruising waters.

MATERIALS OF CONSTRUCTION

Ten years ago, all but a handful of FRP sailboats in the 25- to 45-foot range were constructed using conventional fiber glass reinforcement materials and polyester resins. Most of their hulls were solid FRP laminates. Today, most boats still are constructed using conventional fiber glass reinforcement and polyester resins, but the technology for putting this reinforcement and resin together to produce a boat has changed significantly. If one looks only at boats above 30 feet made in the United States, the number made today using the conventional FRP technology of the 1970s may well be less than 50 percent.

Today, for example, builders must choose between three basic kinds of polyester laminating resins. Though it would be unusual, they may also select an epoxy resin or one of the newer vinylester resins. Fiber glass reinforcing fabrics are available not only in the now traditional nonwoven mat made of short fibers and the woven material made using continuous filaments, but also as one-, two-, or three-layered nonwoven continuous filament materials. Builders can also choose between two types of glass fiber (E-glass or S-glass) as well as such newer reinforcing materials as carbon (graphite) fiber and Kevlar aramid fiber. Increasingly too, builders are looking at alternatives to the traditional solid FRP hull. Boat decks have been cored composite structures for many years because of the need to make them feel solid underfoot. The same kind of structure is enabling builders to pare pounds and, sometimes, dollars from their hulls while preserving or even increasing hull strength.

The beneficiary of all of these developments often is the boat buyer. However, the spreading use of these various materials makes it worthwhile to learn a little bit about each material so that you can judge the importance of a builder's claims for the materials that go into his boats.

Resins

Most boats today contain two fundamental resin types—*gel coat resins,* which provide a thin, shiny outer coating to protect the laminate, and *laminating resins,* which form the structural matrix with the reinforcing fibers.

Gel coat resins: The gel coat on a boat has three basic functions. One is cosmetic—it provides the pretty, shiny surface. The other two are functional: The gel coat resists migration of water into the underlying laminate; and it protects the laminate from ultraviolet (UV) degradation caused by exposure to sunlight. All of this is accomplished by a resin layer that is only about 20 mils thick.

The gel coat resins used by the industry today are generally premium-grade polyester resins. There are, however, two types of polyester gel coat resins. Those better suited for use in the boating industry are made with the chemical "isophthalic acid" for increased resistance to water migration. The gel coat resins which many believe offer the best protection against water migration, resistance to weathering, and toughness, however, also contain the chemical "neopentyl glycol," abbreviated NPG. These resins are commonly referred to as "isophthalic NPG gel coat resins," and builders who use them frequently promote that fact.

The alternative to the isophthalic gel coat resins are those made using "orthophthalic acid." Though these resins have been used by some builders in the past and may still be used today because of their lower cost, "orthophthalic" gel coat resins are not generally recommended for boat construction. They are less resistant to water migration and their use is thought by some observers to be one of the causes of gel coat blistering on a number of boats built in the late 1970s and early '80s when the dramatic increase in the price of crude oil led to equally dramatic increases in the cost of polyester resins.

Laminating resins: Resins used to lay up (or to "laminate") virtually any FRP structure must be formulated to soak into the reinforcing material and form a tough plastic matrix around the individual reinforcing fibers. It must also form a strong chemical bond to both the gel coat "skin" and any subsequent layers of the laminate so that the entire laminate becomes a unified structure. Resins used for boat construction also have at least two other requirements: they must be resilient—tough enough to withstand the constant flexing a boat undergoes through years of use; and they must be easy to use under a wide variety of working conditions.

Most polyester laminating resins used in boat building are orthophthalic resins. They are generally recognized as good general purpose resins for the marine industry. As in the case of gel coat, the substitution of isophthalic acid for the orthophthalic acid component will provide a somewhat tougher resin which also has increased water and chemical resistance. However, that small improvement in resin performance comes with a 10 percent increase in resin cost. While there is a consensus that the added cost is well worthwhile for the gel coat, there is not any such common view in the case of laminating resins: the added cost is not generally thought worthwhile. We need to point out, however, that accelerated aging tests are claimed by the major producer of isophthalic acid to show that use of isophthalic laminating resins can prevent gel coat blistering. If this claim proves true, the added cost for isophthalic laminating resins may well be worthwhile—at least for the resin used in the laminates placed next to the gel coat. But at present, the validity of the accelerated aging tests is subject to question.

A third basic type of polyester laminating resin that builders can use to reduce their costs is produced by substituting a chemical called DCPD for the orthophthalic acid component of the resin. The DCPD laminating resins offer other advantages to builders as well as lower resin costs: They cure more quickly, thereby speeding the laminating process by reducing the time between laminates; and they shrink less, thereby easing the problem of pattern show-through, a problem which arises when the weave of a reinforcing fabric "shows through" the gel coat. At the same time, DCPD is not used without some compromise. The reduced shrinkage may make it more difficult to remove hulls from the molds. In addition, DCPD resins are generally less resilient, less water resistant, and do not soak into the layer of reinforcing fibers as easily as conventional orthophthalic laminating resins.

Epoxy resins are seldom used in production boat building because their cost may be three to four times that of polyester resins. They also are more difficult to use than conventional polyester laminating resins and are not, therefore, well suited to the unskilled labor commonly employed in the lay-up shops of production boat builders. However, the added cost of the resin and the need for more experienced workers pay dividends in terms of performance for those willing to pay for it, generally racers. Epoxy resins are stronger, tougher, and will adhere to almost anything better than polyester resins. These advantages are generally worth their cost in building stronger, *lightweight* boats where the weight savings is particularly important—for example, in winning races. The added cost of epoxy resins usually is not considered worthwhile in building cruising boats, except possibly for bonding components to the interior of a hull.

Vinylester resins are becoming increasingly popular in building sailboats for racing purposes, though it is difficult today to justify their added cost in a boat intended for cruising. These resins are a hybrid, as it were, between polyester and epoxy resins. Compared to polyester resins, they offer improved adhesion, a tougher, less brittle structure, and an ability to withstand greater strain before failure—a combination of properties which makes vinylester resins particularly suitable for use with high strength reinforcement fibers (as opposed to conventional fiber glass materials). At the same time, compared to expoy resins, the vinylesters are easier to use and, at only 1.5-times the price of polyester resins, much less costly.

Reinforcing Materials

Traditionally—even popularly—FRP boats have been and are today "fiber glass boats." The reinforcing materials used are made of glass fibers and they are commonly used in three basic forms—a short, chopped fiber; a loose feltlike fabric made of short fibers; and a heavy, loosely woven fabric made of relatively thick yarns of continuous glass fibers. In the past several years, however, both higher strength fibers and new, nonwoven fabrics made of conventional fiber glass as well as the newer high-strength fibers have been introduced to boat building. As one might expect, these newer materials first found application in boats intended for racing, where performance rather than cost is often the governing factor. More recently, some builders have begun using the new materials in larger (over 40 feet) sailboats intended principally for cruising. The likelihood is that use of these materials will continue to spread.

Conventional fiber glass reinforcing materials: An estimated nine out of ten FRP boats built today are constructed using conventional E-glass reinforcing materials. These are the materials the boating industry has cut its FRP teeth on.

Woven roving is a coarsely woven fabric developed in the early 1960s. It replaced a fiber glass cloth resembling burlap in weight. Although laminates made using the cloth were strong, it seems likely that builders wanted a coarser, thicker fabric to speed construction once they started building boats in the 25-foot range and larger. The hulls they were building would have required a large number of cloth and mat layers to reach the desired hull thickness. Woven roving derives its name from the heavy yarn of continuous filaments from which it is woven. This yarn, which resembles old-fashioned baling twine in appearance, is called "roving"—hence, the "woven roving" terminology.

Several characteristics of woven roving fabrics are worth noting:
• The greatest strength of the fabric runs along two axes at right

angles to each other because the woven fibers are oriented along the length and width of the fabric.
• All woven fabrics have relatively less strength along the diagonal (at 45 degrees to the main axes) because of the right-angle orientation of the fibers.
• The fabric provides greater strength along its length than across its width because about 55 percent of the reinforcing fiber runs lengthwise and 45 percent crosswise.
• Under high stress, the over-under configuration of the woven fabric may tend to break the relatively brittle glass fibers. While this is not a problem in a properly engineered laminate, it does make use of woven roving in very lightweight hull structures more difficult.

Woven roving is generally available in three weights—24, 18, and 40 ounces per square yard. The 24-ounce fabric is by far the most widely used of the three. In contrast, 18-ounce woven roving is used by relatively few builders, in part, at least, because more laminations (and, therefore, more time and labor) are required to build up the desired laminate thickness. However, builders who use the 18-ounce woven roving generally claim they are able to build stronger laminates either because they obtain better glass-to-resin ratios, the resin saturates the fabric more effectively, or both. The basic fact behind this claim is that it is easier to wet out the lighter fabric with resin. Because of its stiffness and the difficulty in wetting it out, the 40-ounce fabric is used principally where extra reinforcement is needed and the builder wants to provide the added strength with fewer layers. An example might be its use in reinforcing the keel area to carry the heavy ballast loads.

Chopped strand mat is a loose feltlike material made of short fibers laid down on a sheet and passed through rollers to press the fibers together. The fibers are held in place by a chemical binder. The nature of the manufacturing process results in variations in the weight of the mat of as much as 20 percent from standard in either direction. Mat generally serves three basic functions in the laminate: (1) It is used as a "skin coat" or backing to the gel coat to ensure a good bond between the laminate and the gel coat resin by providing a relatively smooth bonding surface; (2) it helps provide a strong bond between layers of woven roving by smoothing away the peaks and valleys of the woven fabric's surface; and (3) it contributes to the structural integrity of the laminate by adding strength in all directions and by building up laminate thickness. An advantage of mat as a reinforcing material is that it provides essentially equal strength in all directions because of the random orientation of the fibers. A disadvantage of mat is that it requires an excess of resin to saturate the fabric. The result is added weight and reduced strength because the glass-to-resin ratio—generally in the 25 to 30 percent range—is relatively low. This compares to a glass-to-resin ratio ranging from 40 to 55 percent in a laminate using only woven roving, and from 30 to 45 percent in a laminate of alternating layers of 1.5-oz. mat and 24-oz. woven roving.

For all practical purposes, chopped strand mat is always made of fiber glass. It is generally available in one of two weights—1.5-ounces per square foot, and ¾ of an ounce per square foot. Most builders use the heavier weight mat because fewer laminations are needed to build up desired thickness and the 1.5-ounce mat works well in combination with the commonly used 24-oz. woven roving. Use of ¾-ounce mat, however, may result in a laminate with a slightly higher glass content (and, therefore, greater strength) because the less dense fabric is easier to wet out. In addition, ¾-ounce mat is often preferred for use with 18-oz. woven roving and with the newer nonwoven reinforcing materials.

Chopped roving, more commonly called just plain "chop," consists of short lengths of glass fiber that are essentially "sprayed" onto the laminate with a device called a chopper gun. The "gun" is used to spray both resin and fiber onto the laminate. Fibers are fed to the gun from large spools of roving; they are cut ("chopped") into short lengths by a device on the gun; and, finally, the cut fibers are caught in the spray of resin and "sprayed" onto the laminate (Photo 8). It is literally quick and dirty (messy). Done well, it can also be an effective way to apply the "mat" layer—in this case, a "chop" layer—of the laminate.

The advantage of using the chopper gun is one of cost. Roving (on spools) is the least expensive form of glass-reinforcing fiber. In addition, using the chopper gun is considerably faster than laying up a mat fabric by hand. As a result, the cost savings

8. The chopper gun is used to spray short (about 2-inch) fibers and resin onto the mold. Other workers follow the chopper gun with grooved metal rollers to compact the fibers and work the resin into the laminate.

to the builder can be significant—particularly for large-volume builders. The benefit does not stop with the builder, however, because he will often pass part of the cost savings on to consumers as he competes for their dollars. However, the use of a chopper gun has disadvantages as well. Its principal shortcomings lie with the skill and mood of the operator. The resin usually contains a small amount of dye to help the operator see how much he is applying and where. Moreover, by turning the chopping device on and off as he sprays the resin, the operator can control the relative amounts of glass fiber and resin. As a result, a skilled operator can achieve a glass-to-resin ratio within roughly the same limits achieved using chopped strand mat—about 25 to 30 percent. However, if the operator is daydreaming, is angry with his boss, had a fight with his wife, or is otherwise distracted, all semblance of quality control goes out the window. The most likely result is an excess of resin and glass-to-resin ratios more nearly 20 percent than 30, i.e., a more brittle laminate.

Fabmat and Bi-Ply are trademarks for two products which put woven roving and chopped strand mat together in a single fabric. They are used principally to save time in the lay-up process because the two fabrics are put down at the same time. Most often, they are used in locations in which added reinforcement is needed (as opposed to being used to lay up an entire hull) and for bonding bulkheads and other interior components to the hull. The savings in time, however, is at least partially offset by the higher cost of the two-part reinforcement fabric compared to the cost of mat and woven roving purchased separately. In addition, these materials are much harder to wet out because of their thickness than are mat and woven roving when laid up separately.

Boat cloth, a relatively lightweight woven fabric, is the predecessor to woven roving as mentioned above. While boat cloth is still used in some small boat laminations, its principal application in larger boats is in fabricating such interior components as shower stalls and fuel or water tanks. Use of boat cloth in the laminate provides required strength with less weight than is possible using an all-mat or all-chop lay-up schedule. It is also usually possible to obtain a better glass-to-resin ratio using boat cloth because the lighter weight fabric is more easily wet out.

Modern reinforcing fabrics: The push in the aerospace industry for high strength, lightweight materials has led to development of nonwoven, continuous filament fabrics in which most or all of the fibers run in the same direction. The most commonly used of these new fabrics is "unidirectional" roving. However, "bidirectional" and even "triaxial" nonwoven rovings also are available. These fabrics make it comparatively easy for engineers to orient the reinforcement fibers precisely along the calculated stress patterns. Use of these fabrics also can yield significant weight savings: For example, the smooth surface of a unidirectional fabric (compared to the over-under pattern of a woven fabric) allows use of lighter weight mat between the layers of roving, reducing the weight of the laminate. It is even possible for skilled laminators to eliminate mat layers completely, reducing weight even further. In addition, the ease of wetting out unidirectional fabric and reduction in mat content leads to generally higher strength-to-weight ratios because more of the laminate is reinforcing fiber and less of it is resin.

Unidirectional roving has some 90 percent or more of the fibers in the fabric all running parallel to each other. The fibers usually are held together either by stitching across the fabric at frequent intervals or by using an adhesive to glue a narrow band of fibers across the fabric, also at frequent intervals.

Bidirectional and triaxial rovings are fabrics in which two or three layers of unidirectional rovings are placed one on top of the other and stitched together, most likely using a special knitting machine. In principal, the fibers in each layer can be oriented all in the same or in three different directions. The principal disadvantage of the bidirectional and triaxial rovings is the stiffness of the fabric, making it more difficult to mold the fabric to smaller radius curves. They are also more difficult to wet out than are unidirectional fabrics. Their attraction, on the other hand, is the same as that of Fabmat or Bi-Ply—time saved by laying down two or three layers of fabric in one operation.

Modern reinforcing fibers: Development of high-strength organic and inorganic fibers has added still another dimension to FRP boat construction. None of these fibers was developed for use in the boating industry. All have seen their application in composite (FRP) construction advanced by the high value placed on weight savings in the aerospace industry. As that technology has been developed, however, the companies who sell these fibers have sought to expand their markets into the boating industry. At the same time, people who design and build boats for highly competitive professional and quasi-professional yacht racing have been eager to provide the needed test market to demonstrate the value of these materials in boat construction.

S-glass is a high performance glass fiber providing from 30 to 40 percent higher tensile strength, impact strength, and flexural strength on an equal weight basis than the E-glass fiber used in conventional FRP boat construction. However, the grade of S-glass used in the boating industry is three to four times as costly as conventional fiber glass materials. S-glass normally is used in unidirectional rovings to obtain maximum advantage of its greater strength; it also is most often used to reinforce specific areas of the laminate rather than to lay up the entire hull or deck.

Kevlar 49 is an aramid fiber that is chemically similar to nylon, but whose crystalline structure gives it five times the strength of steel on an equal weight basis. Fibers of Kevlar are 50 percent stronger, 80 percent stiffer, and only one-half the weight of conventional glass fibers of the same size. At the same time, however, Kevlar has a relatively low compressive strength—lower than both conventional fiber glass materials and other newer reinforcing fibers. As a result, while Kevlar can be used to provide increased strength and impact resistance in a laminate with a significant reduction in weight, it should not be used alone where it will have to carry heavy compressive loads.

Kevlar is used to produce both woven and unidirectional reinforcing fabrics. However, because of the strength characteristics of the fiber, the woven fabrics used for FRP laminates are much lighter in weight and bulk than fiber glass fabrics. As a result, the fabrics of Kevlar more closely resemble boat cloth in surface texture than they do conventional woven roving. For that reason, most builders use woven fabrics of Kevlar with ¾-ounce mat, further reducing laminate weight compared to the conventional 24-ounce woven roving and 1.5-ounce mat construction. Some builders eliminate the mat except in the skin coat, saving more weight still.

Initially, the price of Kevlar matched the fiber's golden color, with the result that the fiber was generally used mostly in racing boats or as an expensive option for some high speed power boats. In 1984, however, the price of Kevlar for marine applications was reduced significantly and its use has increased as a result. Despite the reduction in price, the principal application of Kevlar is in boats where speed or operating economy under power (as in a planing hull) are important considerations. However, with increased exposure to the market, use of Kevlar probably will expand eventually into boats intended for racing and cruising, where its high strength and low weight can provide the best of both worlds—reduced weight and the same or better impact resistance compared to a fiber glass reinforced hull for the same boat.

Carbon (graphite) fibers are lightweight fibers with exceptional compressive strength and resistance to bending (stiffness). As a result, they are particularly well suited for such specialized applications in the boat industry as fabricating FRP masts for unstayed sailing rigs and for stiffening hull or deck sections. At the same time, however, carbon fiber has only about 85 percent the strength of Kevlar and is relatively brittle. This latter combination of characteristics results in laminates with relatively low impact resistance. For that reason, even if cost were not an important factor, carbon fibers probably would not be used as an overall reinforcing fiber for the hull or deck of a boat.

Core Materials

It is no accident that the notion "thick is strong" was once a piece of conventional wisdom in the FRP boating industry. Because fiber glass laminates are generally quite flexible, they had to be made stiff enough to prevent excessive flexing and eventual failure—a need normally met by simply piling on more FRP layers. It is a practice still used widely in Far East boat factories, and, to a lesser extent, by U.S. builders. There was—and is today—a certain logic to this approach. The strength of an FRP panel, in terms of its resistance to bending or flexing, lies principally in the outer layers, or surfaces, of the laminate. If those surfaces are close together, as in a piece of paper, the structure bends easily. If they are far apart, as they are in corrugated cardboard, the structure is stiffer, or "stronger." The concept is perhaps most easily illustrated with a piece of lumber. If you put a one-inch-thick board across two sawhorses to make a scaffolding from which to work on your boat, the board will probably bend almost to the point of breaking when you climb up on it. Put a two-inch-thick board across the same saw horses and it will support you easily even if it does bend a little. Put a four-inch-thick board across the same sawhorses—if you can lift it—and it will support both you and your sailing partner and still remain "stiff as a board." About the same can be said for a solid FRP laminate. By adding the layers and making it thicker, you make it stiffer, stronger, and heavier.

Core materials offer an alternative to simply adding more

FRP layers for making stiffer and stronger composites. With a relatively thin FRP skin (laminate) on each side of the core material, greater strength can be obtained with relatively little added weight because the core provides the needed thickness rather than weight (Fig. 3).

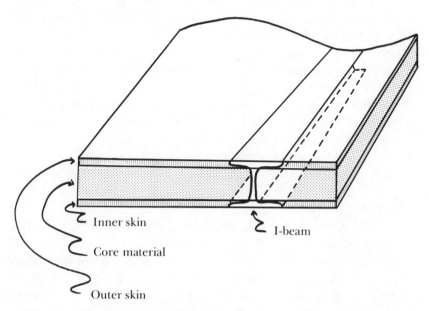

FIG. 3. The core in a laminate serves the same function that is served by the vertical member of an I-beam. It holds the inner and outer (upper and lower) skins a fixed distance apart, thereby increasing resistance to bending.

As noted earlier, decks have long been cored. Because they must carry our weight when we walk on them, deck structures need to carry heavy point load (from 100 to 200 or more pounds on an area the size of a person's foot) or they will not have the solid feel most people want underfoot on a boat. For many years, blocks of mostly ½-inch plywood were widely used as the core material for decks. Today, because of both its high cost and weight, plywood increasingly is being replaced in decks by one of the newer closed-cell foams, end-grain balsa wood core material, and, in a few instances, by a combination mat / microballoon core material. These and other new core materials also

are being used widely today in boat hulls. From the perspective of a cruising boat, a cored hull offers certain advantages:
• The strength and / or impact resistance of the cored construction offers a margin of safety if you are sailing where you may run into heavy floating debris. Even in relatively protected waters, floating debris may include logs or tree trunks unloosed by storms. Offshore, "debris" may be a shipping container lost overboard from a freighter.
• The core material may provide thermal insulation. In the tropics, that insulation may not be as important as in colder climates unless your boat is air conditioned. In colder weather, the insulating effect of the core not only helps keep the boat warmer, it prevents "sweating"—the formation of condensation on the inside of the hull and deck.
• The core material may reduce sound and vibration.
• The use of a core allows the boat's designed displacement to be taken up in tankage and stowage rather than by the weight of a heavy laminate.

In attempting to read through the claims made for the "stiffness" or "strength" of competitive conventional core materials, the nontechnical boat buyer can easily be misled. All of the claims are based on tests of laminates made using each of the three basic core materials—balsa, rigid PVC foam, and rigid-elastic PVC foam—and most of those test results have meaning only to engineers and yacht designers. Most of the time, the designer, or his engineer, can compensate for the different stiffness or strength characteristics of each core material by varying the thickness of the core, the laminate schedule for the FRP skins, and interior reinforcing members. However, the choice of core material can affect interior design, weight distribution, and the ultimate impact strength of the hull.

End-grain balsa is the most widely used core material in the United States today. It is also both the strongest conventional core material by a wide margin and the lowest in cost. End-grain balsa is made by cutting slices from the balsa tree trunk much as a butcher cuts slices from a package of salami. The slices of tree trunk are then cut into small squares and rectangles, which in turn are bonded to a lightweight fabric resembling cheesecloth or mosquito netting in appearance. Sheets of small balsa squares can then be put into a laminate and they will drape to fit the shape of the mold.

Aside from correct specification of the laminate schedule, there are two keys to successful use of end-grain balsa in any laminate. First, the balsa core must be sealed carefully wherever a fastener or piece of hardware passes through the core. Second, each little square of balsa must be bonded securely into the laminate so there are not voids between the FRP skins and the balsa. The bond can be checked by tapping the core lightly before the second skin is applied; any voids will be revealed by a "hollow" sound from the tapping and can be repaired before the second skin is applied.

Both of these " keys" relate to the vulnerability of balsa to water. In principal—and it is a principal which Lloyd's of London has accepted by agreeing to certify boats constructed with a balsa core—water is not allowed to migrate through the core by the end-grain structure of the balsa itself. The grain of the wood, which runs from the outer to the inner skin, prevents water from migrating sideways and thus spreading through the core. The problem, however, is that pathways for migration of water can exist between those little blocks of balsa if resin hasn't filled those cracks. Similarly, a void between one of the skins and the core may provide a route for water migration—if water is allowed to penetrate the FRP skin. And therein lies the importance of sealing the core wherever a hole is made in the laminate. Where they know that fittings will be attached—a through-hull fitting or a deck fitting, for example—many builders either omit the core completely at that spot or substitute plywood for the balsa to isolate the balsa core from the fitting. When holes must be cut or drilled through the balsa core, the edge of the core should be sealed either with gel coat or a first-rate sealant.

Two fundamentally different closed-cell PVC foams are marketed as core materials for boat construction. One, sold under the trade name Airex foam, is a "pure" PVC foam that is rigid-elastic in nature. What this means is that the foam is stiff up to a point, and then it bends. The other PVC foam product is best known in the United States by the tradename Klegecell. (The trade names Divinycel and Plasticell are used for identical foams made in Sweden and the United Kingdom, respectively.) It is a mixture of PVC and polyurethane-type components that results in a "rigid" foam. Although Klegecell is 30 to 40 percent stiffer than Airex, it breaks rather than bend when the forces become too great. Both types of foam cores are good products, but each offers clearly identifiable advantages.

Klegecell foam, a strong, rigid, but somewhat brittle foam, has been widely used in boat construction for many years. It is about one-third less expensive than Airex foam and offers builders the highest strength-to-weight ratio of any conventional core material. Compared to a similar laminate using an end-grain balsa core, for example, Klegecell offers somewhat *more than* half the strength at somewhat *less than* one-half the weight. The weight difference results from two factors: (1) The Klegecell foam itself weighs only half as much as the balsa core material; and (2) in use, the end-grain balsa soaks up a certain amount of resin, whereas the foam does not. The strength-to-weight comparison with Airex foam is more dramatic. In tests of comparable laminates, the Klegecell composite is 30 to 40 percent stiffer than the Airex panel. At the same time, the Airex foam is four times the weight of the Klegecell core. This high strength-to-weight ratio makes Klegecell composites particularly attractive for use in boats designed for racing and in deck and cabin house structures.

The chemical structure of Airex foam is noteworthy for the absence of any chemical crosslinking. As a result, Airex foam can be bent around an inch-thick bar to a U-shape; it can also be compressed to about one half of its original thickness without rupturing the cells which provide the foam structure. Afterward, the foam will return slowly to its original dimensions. The practical effect of this flexible characteristic is twofold:

• After an initial stiffness, a laminate made with an Airex core will bend under pressure. However, the core will not fail as long as at least one of the two skins remains intact. Laminates made using either end-grain balsa or rigid foam core materials, in contrast, are significantly stronger initially, but fail catastrophically when the forces exceed their structural properties.

• Laminates made using Airex as the core material provide significantly greater resistance to damage from impact than either solid FRP laminates, those made with other foams, or those with balsa cores. There is one principal reason for the greater impact resistance of Airex-cored laminates: the Airex core tends to absorb the impact, thereby transmitting comparatively little of the impact energy to the inner skin. In contrast, solid FRP laminates, balsa core, and rigid foam cores tend to pass the impact energy straight through to the inner skin. As a result, the following occurs in impact tests: The solid FRP lay-up delaminates from the force of the impact; the balsa core breaks away from the inner skin, with some delamination occurring in the inner skin; the rigid foam core shears and breaks away from the inner skin, with some delamination occurring in the inner skin; the outer skin of the Airex composite is mildly damaged by the impact, but no other damage is evident.

While the Airex offers an advantage of impact resistance, its flexibility and lower stiffness compared to other core materials demand compensation in the boat structure. For example, a deck made using Airex foam for a 35-foot boat will require either a thicker core or an extra laminate of woven roving and mat in one of the skins for added stiffness. Both alternatives add weight and expense compared to either balsa or a rigid foam core, such as Klegecell. Similarly, a hull made using Airex will require more interior stiffeners than hulls made using a similar laminate schedule and cored with balsa or Klegecell. Finally, because the Airex core can be compressed easily, builders must use care in installing through-hull fittings or other hardware. If the core was not omitted in favor of a solid FRP laminate where fittings will be installed, either a compression sleeve should be used over bolts before fasteners are tightened down, or the core should be removed and a wood insert used to replace it.

Coremat is a loose, white, feltlike polyester material filled with microballoons—dustlike particles that are microscopic-sized hollow plastic spheres. Coremat is exceptionally light in weight and is sometimes used to replace fiber glass materials in the middle of a laminate to reduce the overall weight of the structure. That weight savings comes from two sources: The Coremat soaks up slightly less resin than conventional fiber glass materials for equivalent thickness; and, the microballoons occupy about 50 percent of the volume of the Coremat layer, or layers, of the laminate.

Coremat is attractive to some builders because it is applied in the same way that fiber glass mat is applied, thereby saving the added labor associated with more conventional core materials. In addition, because it is available in thicknesses up to 5 millimeters (almost one-quarter of an inch) and is easy to use, Coremat makes it possible to build thickness more quickly than using multiple layers of conventional fiber glass reinforcement. However, the Coremat material has no structural properties of its own and does not contribute to the structural strength or impact resistance of the laminate. If the Coremat is used to increase the thickness of the laminate, the increased thickness will add to panel stiffness, but the actual strength of the Coremat layers in the laminate is limited to the strength of the resin. As a result, thick layers of Coremat tend to fail under stress much more quickly than laminates made with conventional core materials.

All of the mechanical properties that make plywood an excellent building material make it the strongest of the conventional core material for FRP composites. However, plywood brings with it three major disadvantages in boat building today: It is relatively heavy; it is more vulnerable than any of the other core materials to water damage; and the labor involved in installing a plywood core in any but small areas is prohibitive for most builders. Despite these disadvantages, plywood is still used as a core material. The builder of at least one series of strongly constructed ocean-cruising sailboats, for example, uses a plywood core throughout his deck and cabin house laminate. More commonly, however, plywood is used in place of other core materials where deck hardware will be attached.

FUNDAMENTALS OF FRP CONSTRUCTION

The actual process of laying up an FRP hull or deck is not particularly difficult. The industry has developed a number of good laminating procedures over the years to accommodate the idiosyncracies of the resins and reinforcement fabrics used to make the lamination. These procedures provide guidelines needed to enable builders to do their jobs well. However, the procedures must be followed not only by policy of the builder, but in practice by his workers. And how well the procedures are followed has much to say about the quality of any individual hull or deck. For example:

• Polyester resins cure in two stages. Initially, the resin is a syrupy liquid. Once the chemical initiator—popularly called a "catalyst"—is added to the resin, the cure begins. Depending upon the resin type and the amount of initiator added, the resin may remain syrupy for as little as 10 or 15 minutes, or as long as two hours or more. At some point, the cure accelerates greatly and the syrup first becomes jellylike, then changes to a soft solid, and, finally, to a hard solid. The key point: *Once the resin begins to gel—to change from syrup to jelly in form—it should not be disturbed.*

• Polyester resins often generate a significant amount of heat in the curing cycle; the amount of heat released is related to the specific resin formulation and to the thickness of new laminate. As a result, builders can generally lay up only about two laminations (each lamination consisting of one layer of mat and one of woven roving) per day. The key point: *Several days are needed to complete the lay-up for the hull or deck for many cruising sailboats. For larger boats, well over a week may be needed.*

• Polyester resins may be quite hard to the touch after only a couple of hours of cure time; however, the cure continues for several days. In the first day or two of the curing process, additional laminates can be applied and a strong chemical bond (a primary bond) will be formed between the new layer of resin and the "old." After 72 hours, however, the cure has reached a stage at which the new layer of resin cannot forge an adequate bond with the cured surface without scuff-sanding the laminate before applying the resin. The bond formed after scuff sanding is called a "secondary" bond. The key point: *If an interval of more than two-and-a-half days (Friday p.m. to Monday a.m.) occurs in the laminating schedule, the surface of the laminate should be scuff-sanded before applying a new laminate to permit a strong bond between the old and the new.* Some builders, in fact, recommend scuff-sanding if more than a single day passes between laminations. The same facts apply to the process of bonding interior components to the hull, or to making a hull-to-deck joint: the surface or surfaces to be bonded should be scuff-sanded if they are more than 48 to 72 hours old.

• Fiber glass reinforcement is opaque—white—before the resin is applied. As each fiber becomes wet with resin, it turns transparent. What this means to the laminator is that he can tell when the fiber glass is completely wet because he no longer sees any white fibers. The key point: *You can determine whether a laminate has been adequately wet out simply by looking at it. If you can see the white glass fibers or milky white spots in the laminate, the fabric has not been fully saturated.* The unsaturated fibers do not contribute fully to the strength of the laminate; in addition, they can wick moisture into the laminate if the fiber ends are exposed. In fact, such "voids" or resin-starved areas in the laminate next to the gel coat are another possible explanation of the cause of gel coat blisters; the water is thought to migrate slowly through the gel coat and collect in those voids, establishing an osmotic potential across the gel coat. As more water crosses through the gel coat to balance out the osmotic potential, blisters are formed. In turn, the liquid in the blister is said to attack the resin in the laminate, changing the osmotic balance, drawing more water through the gel coat and enlarging the blister. Such blisters can render a boat unseaworthy if they are allowed to grow far enough into the laminate. We have seen blisters larger than your hand and from three-eighths to a half inch into the hull laminate. If this theory is correct, wetting out the skin coat and subsequent laminates properly takes on added importance.

• In the lay-up procedure, a coating of resin should be applied before the fabric is put down. In this way, the resin is able to soak into the fabric from the bottom up as workers roll the laminate using grooved metal rollers. When the fabric has been rolled adequately, additional resin can be applied as needed to complete the saturation process. The key point: *This two step process makes it easier to avoid using an excess of resin by making it easier to saturate the fabric.* Often, when a layer of mat or chopped roving has been put down, the layer of woven roving is applied before the resin in the mat layer has gelled. Because the mat layer normally has an excess of resin (70 to 75 percent resin by weight), the woven roving often can be wet out from the bottom by rolling the roving firmly into the mat layer.

• The lazy-man's way to obtain complete saturation of the reinforcing fabric is to use an excess of resin. However, laminates in which there is an excess of resin tend to be brittle. The key point: *A laminate in which there is an excess of resin will tend more toward a glassy appearance. By contrast, the texture of the reinforcing fabric or fiber is quite distinctive in both touch and appearance in a laminate with a good resin / glass ratio.*

This list is not all-inclusive, but it is sufficient to illustrate the central point: Certain of the good laminating procedures are the responsibility of the builder—for example, requiring scuff-sanding if too long an interval passes between laminations; responsibility for following most of these procedures, however, falls on the laminators themselves and their supervision. Working in the laminating shop is an entry level job in many boat factories. Moreover, the laminator's job is not generally a pleasant one. Fumes from the resin are quite pungent and, even with excellent ventilating systems (which many builders do not provide), the workers stand in the midst of and are constantly bending over fresh resin, breathing in the vapors. That is the environment in which your boat hull and deck will be constructed; and it is that environment which makes the quality of supervision and the workers' pride essential determinants of the quality of the laminate.

From your viewpoint as a boat buyer who may visit one or more builders before reaching your final decision, one other point is worth noting: Although working with FRP materials can be a messy occupation, particularly when a chopper gun is used, many builders are able to maintain very neat and clean laminating operations. Our own feeling is that a builder who maintains a clean, neat laminating operation demonstrates a pride and attention to detail that is likely to be contagious throughout his employee force and to result in a well-constructed hull and deck.

Solid Laminates

Solid FRP laminates are more often used in hulls rather than decks, and to construct smaller components such as tanks and interior liners. In addition, most solid laminates are made using conventional FRP materials; the level of technology represented by high strength fibers and unidirectional rovings is more consistent with cored construction.

In all but a small number of custom boats, solid FRP laminates are made using a female mold. One common example of a female mold is a plastic ice-cube tray. Two others are cake pans and Jello molds. A hull mold is similar in concept. It provides the shape and surface characteristics to the outside of the hull—just as an ice-cube tray, a cake pan, or a Jello mold provides the shape and surface characteristics of the ice cube, cake, and Jello.

Most boat hulls in the United States are designed for production using one-piece molds. It is less costly because the molds do not have to be assembled and disassembled. In addition, either the buildings at the factory are tall enough to lift the hull from the mold, or the mold can be wheeled outside, the hull removed, and both the mold and hull moved back under shelter. Sometimes, however, either the boat design or space limitations at the factory dictate use of molds consisting of two or more pieces that are bolted together when laying up the hull and taken apart to remove the hull. In the case of a two-piece mold split down the middle along the keel, builders often lay up the initial layers of the laminate for each side of the hull separately, bolt the two halves of the mold together, and join the two halves of the laminate by laying progressively wider layers of reinforcement along the centerline, overlapping the hull pieces. This procedure leaves a groove down the centerline of the boat on the outside of the hull that is filled in later with a polyester putty and usually is difficult to detect. Correctly done—and that includes design of the laminate schedule as well as the actual laminating process itself—a hull laminated in two halves and subsequently bonded together is theoretically as strong as a comparable hull in which the two sides are laminated as a single piece. However, for offshore sailing and with due regard to theory, we would prefer a hull laminated as a single unit, with all layers of reinforcing fabric, including the outermost layers, bridging the centerline of the boat.

In some instances, two-piece molds are treated as a single mold from the beginning of the lamination process, but the two-piece mold design is required because of hull shape. In this case, although the two pieces of the mold are separated to remove the hull, the hull is laminated as a single unit. Design features requiring a two-piece mold include tumblehome and a reverse transom.

Even molds in which the two halves of the hull are inseparable may have two or more removable parts. For example, a reverse transom in a "one-piece hull mold" normally requires a separate mold panel which can be removed before the hull can be pulled. Similarly, boats with internal hull flanges—flanges along the sheer that turn inward and serve as the platform for fastening down the deck—require separate mold strips along the sheer. The mold strips for the internal hull flange must be removed before the hull can be pulled. They also must be put back into place before the next hull is made. The bottom portion of a deep keel or a skeg also may require separate pieces for the mold because of difficulty reaching those areas to lay up the reinforcing fabric.

The inner surface of a hull mold normally is highly polished. This polished surface is what gives the gel coat its smooth surface and gloss. Before a hull is laid up, the mold is coated with a waxlike material called a mold release agent. Often, a single good application of the mold release agent will last through production of a dozen or more hulls. The more complex the shape, however, the smaller the number of lay-ups that can be done between applications of the mold release agent.

The first step in laying up the hull is applying the gel coat to a nominal thickness of 20/1,000 of an inch. Much thicker and gel coat cracking or crazing may be a problem. Much thinner and the gel coat can't do its job of protecting the laminate adequately. The gel coat is applied with a spray gun by fogging it onto the mold surface to achieve the desired thickness in a single, continuous application.

When the gel coat has fully cured (the cure should be tested using a spring-loaded device with a sharp pin that is pushed into the gel coat) the initial layers of the laminate are applied. Normally, the "skin coat"—the part of the laminate adjacent to the gel coat—consists of three ounces per square foot of chopped strand mat or chopped roving. The only advantage to use of chopped roving is cost. Roving costs less than mat. Moreover, chop can be applied quickly using a chopper gun, thereby providing cost savings in labor compared to putting down mat by hand. The mat first must be cut to size. Then a coating of resin is applied to the mold, the mat laid out onto the resin, more resin applied, and the resin worked fully into the mat to saturate the reinforcing fibers. (Unlike fabrics made of roving, mat cannot be rolled into the underlying resin layer without being wet from the top because the roller tends to break up the mat surface if it is dry.) In using the chopper gun, the gun operator sprays some resin over an area, sprays a mixture of chopped fiber and resin, and moves onto the next area while fellow workers roll out the chop behind him. In any case, the skin coat usually is applied in two passes—two layers of 1.5 ounces of mat or chop, with the second layer applied right after the first layer has set up hard. As noted earlier, complete saturation and thorough rolling out are particularly important in these first layers of mat or chop to prevent voids, or air pockets, from forming beneath the gel coat.

The skin coat serves only a limited structural function, serving mostly to provide a transition between the smooth inner surface of the gel coat and the wafflelike surface of the first layer of woven roving. Its second role is principally cosmetic—that is, it hides the "waffle" pattern of the woven fabric commonly used in the remainder of the laminate. (In boats constructed of unidirectional roving, the skin coat may be somewhat thinner because there is no waffle pattern to show through.) Subsequent laminations provide most of the structural properties of the finished hull. Usually, additional laminates are applied in pairs—one layer of chopped roving or mat followed by a layer of woven roving (Table 1).

A few builders, particularly in the Pacific Northwest, lay multiple layers of woven roving right on each other, doing away with the mat interlayers. Done well, this can yield a strong laminate with a relatively high glass-to-resin ratio. The difficulty, however, is in getting a good bond between the laminates. If the first layer of woven roving is allowed to cure before the subsequent layer is applied, the bond between the first and second layer may be questionable because of the difficulty in get-

TABLE 1. Typical Lay-up Schedules for Solid FRP Hulls
(These examples do not include areas of extra reinforcement, e.g.,
around chain plates)

	Topsides	*Bottom*	*Keel*
30' racer	Gel coat	Gel coat	Gel coat
	1.5-oz. mat	1.5-oz. mat	1.5-oz. mat
	1.5-oz. mat	1.5-oz. mat	1.5-oz. mat
	24-oz. woven roving (WR)	24-oz. WR	24-oz. WR
	1.5-oz. mat	1.5-oz. mat	1.5-oz. mat
	24-oz. WR	24-oz. WR	24-oz. WR
			1.5-oz. mat
			24-oz. WR
32' cruiser	Gel coat	Gel coat	Gel coat
	1.5-oz. mat	1.5-oz. mat	1.5-oz. mat
	1.5-oz. mat	1.5-oz. mat	1.5-oz. mat
	24-oz. WR	24-oz. WR	24-oz. WR
	1.5-oz. mat	1.5-oz. mat	1.5-oz. mat
	24-oz. WR	24-oz. WR	24-oz. WR
	1.5-oz. mat	1.5-oz. mat	1.5-oz. mat
	24-oz. WR	24-oz. WR	24-oz. WR
		1.5-oz. mat	1.5-oz. mat
		24-oz. WR	24-oz. WR
		1.5-oz. mat	1.5-oz. mat
		24-oz. WR	24-oz. WR
			1.5-oz. mat
			24-oz. WR
			1.5-oz. mat
			24-oz. WR
35' racer/cruiser	Gel Coat	Gel Coat	Gel Coat
	3-oz. mat	3-oz. mat	3-oz. mat
	1.5-oz. mat	1.5-oz. mat	1.5-oz. mat
	24-oz. WR	24-oz. WR	24-oz. WR
	1.5-oz. mat	1.5-oz. mat	1.5-oz. mat
	24-oz. WR	24-oz. WR	24-oz. WR
		1.5-oz. mat	1.5-oz. mat
		24-oz. WR	24-oz. WR
			1.5-oz. mat
			24-oz. WR

ting the second layer of woven roving to conform to the hardened waffle surface of the first layer. By using a slow-curing resin so that each subsequent layer of woven roving is put down into a still wet layer behind it, however, the fibers in both fabric surfaces can be moved around slightly as the resin and fabrics are worked with rollers so that the two surfaces are made to conform to each other and provide a better bond.

The same technique is often used even in laying up alternating layers of woven roving and mat. The resin is catalyzed to remain liquid long enough for the roving to be applied before the resin in the mat layer has begun to gel. In this way, two objectives are served: (1) a better mating surface is formed between the mat and roving because the mat conforms somewhat to the waffle pattern of the woven fabric as the roving is rolled into the mat surface; and (2) glass-to-resin ratios are improved because the roving is wet out from the bottom by the resin in the mat, in effect removing excess resin from the mat layer. Although the last—or innermost—layer in the laminate is usually woven roving, a few builders apply a final lightweight layer of mat. They offer two reasons: one is cosmetic ("It looks nicer"); the other is more functional: the smooth mat surface is easier to finish off with gel coat in the bilges to protect the laminate from bilge water.

Cored Laminates

Although cores have been used in deck laminates for years, builders today increasingly are using cored laminates for sailboat hulls in the over 25-foot range. In some instances, the motivation may be economic: With the use of an end-grain balsa core, for example, builders may be able to reduce their costs in producing a hull, particularly in larger boats. Another motivation for the use of cored laminates may be use of interior space. The inherent stiffness of some cored construction allows builders to open up interior accomodations, increasing the sense of space. And still another reason for using cored laminates is the potential for improved performance—and by that, we mean sailing performance. The use of a cored laminate can lead to reductions in the hull weight of up to 50 percent compared to solid FRP laminates of the same strength. This weight savings is accomplished because the core replaces some layers of reinforcing fiber and the resin used with them. Deck weight also can be reduced by using balsa or foam cores in place of the heavier plywood. This kind of weight reduction can have a domino effect. Designed displacements can be lower, allowing smaller sail plans, lighter weight standing rigging, smaller engines, etc.—all of which can translate into lower costs and more competitive pricing for the finished boat. Assuming good design, lighter weight may also be translated into improved sailing performance, particularly in light air, and the improved sailing performance in turn can be used as a marketing plus. In boats intended for long-distance cruising, there may be other ways to use the weight savings in hull and deck laminates to good advantage. For example, a few hundred pounds in reduced hull weight can be used to increase the boat's dry stowage capacity, or the amount of fuel and water it can carry.

However, use of a core adds several extra steps to the laminating process. In fact, builders estimate that laying down the core involves about 25 percent more labor than application of a typical 1.5-oz. mat / 24-oz. woven roving laminate. For example, a typical lamination schedule for a 30- to 35- foot cored racer / cruiser sailboat hull may call for two layers of 1.5-oz. mat to back up the gel coat and then a lamination of 24-oz. woven roving and 1.5-oz. mat to complete the outer skin and a lamination of 1.5-oz. mat and 24-oz. woven roving as the inner skin of the sandwich (Fig. 4). The core is placed between the last mat layer of the outer skin and the first mat layer of the inner skin.

In the case of end-grain balsa, the core is pressed down into a resin-rich mat layer until resin is squeezed up between the blocks of balsa. After a few minutes, it is pressed into the mat again to ensure good adhesion. When the laminate is cured, workers are supposed to test for voids by tapping the core. (A void is indicated by a hollow sound.) Any voids should be repaired before the inner skin is applied. Repairing voids at this stage of the hull construction is particularly important with balsa to minimize any possibility of water entering and damaging the core. After checking for and, if necessary, repairing any

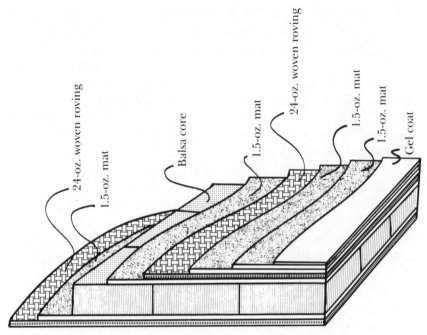

Fig. 4. Typical hull lay-up for a 30- to 32-foot racer/cruiser constructed using an end-grain balsa core.

voids, the inner skin is applied. First the balsa core is coated with resin; the mat is put down; and thereafter, standard laminating procedures are followed until the laminate is completed.

The procedures for installing Airex and Klegecell foam cores are similar. There are, however, two principal differences. First, the last mat layer of the outer skin need not be particularly resin rich. Although the resin will be squeezed up between the small squares of the core, the foams do not soak up resin as does the end-grain balsa. Second, tapping the foam core will not reveal voids. Instead, scraping the foam with a coin or some such small flat disc or square will reveal any voids by a change in the sound. Although voids between the skin and a foam core do not present any risk from water damage to the foam core, these voids should be repaired because the strength and impact resistance of the hull depend upon a good bond between the FRP skins and the core material.

One other note: In a cored laminate, the core material should be continuous from bow to stern. If the core is removed in a line down the side of the hull, the hull may flex around the area where there is no core because the thinner, solid FRP area is not as stiff as the cored laminate. This means, for example, that interior bulkheads should be attached to the inner skin of the sandwich. The exception would be small areas where the core is removed to accommodate fittings or hardware that will be bolted to the hull—for example, chain plates.

Interior Installations

The hull and deck of a boat form an outer shell whose principal structural function is to keep all water on the outside. By itself, however, this shell has relatively little strength. True, the use of cored construction provides a certain level of strength. However, in all but some exotic, lightweight, and relatively small racing boats, an interior framework is needed to help the outer shell retain its structural and watertight integrity. Moreover, the more demanding the sea conditions the boat will be exposed to, the more important the interior framework becomes.

Some parts of this interior framework carry the compressive loads of the rigging. Other parts stiffen the bottom and sides against the forces of the water as the boat rushes from wave to wave. When hulls were uniformly heavy, most of this reinforcing framework consisted of the interior furnishings—the cabin sole and its supporting floor members, bulkheads, and furnishings themselves. As builders have turned to lighter-weight constructions—thinner hull sections—sailboat designers have engineered systems of longitudinal stringers (frame pieces running parallel to the keel) and transverse frames to stiffen the bottom sections of the hull.

From the cabin sole up, three basic approaches to putting the interior into an FRP boat have been developed. One is the traditional approach of bonding all interior components to the hull with FRP laminates. The second approach involves molding most of the interior furnishings and framework into FRP hull and deck liners. The third approach blends the first two: an FRP pan or partial hull liner serves as the base for fabricating the interior. When the interior assembly has been completed, the entire structure is placed into the hull as a unit and fixed in place, bonding bulkheads and other components to the hull with FRP laminates. It goes almost without saying that each approach offers advantages. It is perhaps more important to note that each also has limitations, some of which may be substantial.

The traditional approach: This is a relatively labor intensive and, therefore, expensive way to build a boat. Usually, the main structural bulkheads are the first pieces installed—even before the hull has been removed from the mold. This helps ensure that the hull retains its intended shape. In addition to those principal bulkheads, each piece of the interior is put into place individually and those which join the hull are bonded to it with FRP laminates. The process is called "taping" the bulkheads (or shelves, etc.) to the hull. Subsequently, all of the bulkheads which extend to the overhead are (or should be) taped to the underside of the deck and cabin house so that the entire hull-and-deck structure is tied together around its cross-sectional circumference as well as around the hull-to-deck joint. In addition, the cabin sole and all horizontal furniture surfaces which extend to the hull should be taped to the hull with at least one laminate of mat and woven roving, preferably with two such laminates in boats above 30 feet.

By "taping" or "bonding," we mean using fiber glass mat, woven roving, and either polyester or epoxy resin to affix the component to the hull. As in laying up a hull or deck, there are a number of generally accepted (though too often ignored) good laminating procedures for taping interior components to the hull or deck. They include the following:

• The hull and deck should be wiped free of dust using a solvent and the surface scuff-sanded to ensure a good bond between the inner hull surface and the laminate used to tape the component to the hull.

• All bulkheads should have a spacer between the edge of the bulkhead and the hull. Some builders use a caulk or puttylike compound to form that "spacer." The much preferred approach uses a specially shaped piece of foam as the spacer. The foam is called a fillet, hence the name "fillet bond" for that kind of joint. Looked at from the end, the fillet is shaped like a triangle

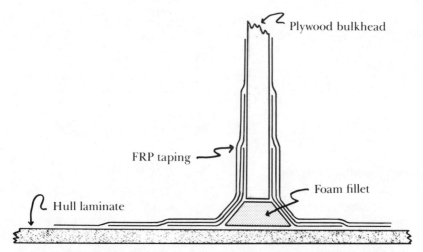

FIG. 5. A hat-section foam spacer—called a fillet—provides a gentle radius for the FRP taping to eliminate weak spots. The fillet also helps transmit impacts on the hull directly to the wood bulkhead, reducing stresses on the bond itself.

with its top sliced off so that it forms a gentle radius from the bulkhead to the hull (Fig. 5). The width of the fillet at the hull should be two to three times the thickness of the bulkhead. With the fillet (or putty spacer) in place, the bulkhead is taped to the hull on both sides of the joint. The fillet bond has three principal functions: (1) It helps prevent the bulkhead (or other panel) from creating a hard spot in the hull which may subsequently become a focal point for flexing of a hull panel; (2) it provides a gentle radius for the taping used to bond the bulkhead to the hull; and (3) it prevents the laminates from folding back between the edge of the bulkhead and hull, where they can be broken (Fig. 6). Fillets should also be used for bonding any other vertical panel to the hull—e.g., sides of hanging lockers and galley or head cabinets. In fact, there is a rule of thumb: "If it touches the hull, tape it."
• The FRP laminates bonding wood components to the hulls should extend a full six inches onto the wood and from four to six inches onto the hull. The larger overlap onto the wood is necessary because the polyester resin does not adhere strongly to wood. If the bulkhead or other component has a veneer—

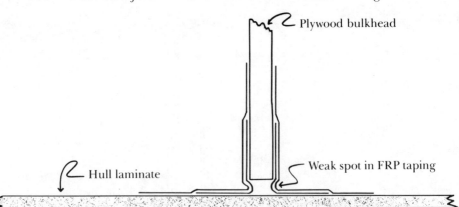

FIG. 6. When no fillet or other spacer is used to hold the bulkhead from the hull, the FRP taping often folds into the gap between the edge of the bulkhead and the hull, forming a weak spot. Any impact on the hull—whether from banging against a piling or beating to windward through a heavy chop—will tend to squeeze the folded area and may fracture the FRP tape. Frequently builders who omit fillets use only one—at the most two—layers of FRP taping to make the bond.

particularly a teak veneer—the veneer should be removed before the FRP bond is made to ensure good adhesion between the resin and the wood.

• Although some builders use a single laminate of woven roving to tape panels to the hull, alternating layers of mat and woven roving are recommended, with the mat put next to the hull and the wood surface of the bulkhead. Mat will provide a better bond to both surfaces.

• Two or three laminates are recommended for taping bulkheads and other major vertical panels to the hull, depending upon the size of the boat. We would prefer to see three laminates in any boat over 30 feet in length. Typically, the first laminate (1.5-oz. mat and 24-oz. woven roving, or a similar weight Fabmat) should extend two inches onto both the hull and bulkhead. The second laminate should extend four inches onto the bulkhead and four inches onto the hull. The third laminate should extend the full six inches onto the bulkhead and from four to six inches onto the hull.

• In places where it is impractical or undesirable to extend the FRP bond six inches onto the wood component, the FRP laminates can be treated like an angle iron extending two or, preferably, three inches onto the wood and made fast to the bulkhead with machine screws, lock washers, and nuts.

• Wherever major horizontal interior components—the cabin sole, berths, counters, and cabinet tops—extend to the hull, they should be taped to the hull with FRP laminates extending six inches onto the wood and three to four inches onto the hull surface. Fillets are not necessary. Some builders, however, use an adhesive/sealant to help form the joint between the horizontal component and the hull before applying the FRP bond.

Under no circumstances should a bulkhead or major interior component be set a distance away from the hull and only the FRP tape used to bridge the space between the bulkhead and hull. In such an installation, the relatively brittle FRP laminate must carry the forces from any impact—the boat banging against a dock, or pounding through the waves heeled well over in a stiff upwind beat. The potential for that joint to fail in such circumstances exists from two directions (Fig. 7): The taping itself may fracture; alternatively, the shear forces created by the repeating pounding along the plane of the tape may break the

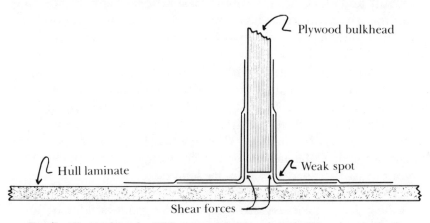

FIG. 7. Even when the FRP tape used to bond the bulkhead to the hull does not fold into the joint as in Fig. 6, any impact on the hull will tend to break the bond unless a fillet or other spacer is used. In this illustration, impacts on the hull create shear forces on the relatively weak adhesive bond between the resin in the FRP taping and the wood surface of the bulkhead. Also, the sharp corners in the tape as it makes the turn from hull to bulkhead create weak spots in the taping itself.

bond between the resin and the wood bulkhead. In a proper installation, the fillet bond absorbs some of the impact loading and transmits the remainder to the bulkhead.

We believe this method of attaching interior components provides the strongest method of building FRP boats. The concept is similar to that of so-called unitized construction in the automobile industry. All of the components contribute to the overall structural integrity of the boat. However, the taping must be done properly, and too often it is not. The most common shortcomings are failure to extend the bond far enough onto the wood bulkhead, failure to remove any veneer from the bulkhead, failure to tape bulkheads to the overhead, and failure to use sufficient layers of reinforcing fabric to provide an adequate laminate (taping) structure. These shortcomings can be hazardous in any sailing where the boat may encounter severe weather. Continued pounding in the heavy, steep chop often found in large coastal or inland bays and the forces involved in heavy ocean wave patterns during a gale have popped bulkheads or broken interior furnishings loose even in relatively well constructed boats.

Hull and deck liners: In the past several years, increasing numbers of production boat builders have begun using hull liners to build the interiors of their boats. Deck liners also have been widely used as an alternative to cored construction to stiffen deck moldings, position and hold bulkheads in place, and to provide an easily cleaned and maintained interior finish. Most often, these interior liners are made entirely of chopped fiber using a chopper gun; rarely do they have more than a single layer of woven roving. Nor should they. Interior liners do not have any particular impact or tensile strength requirements that would justify a sturdier laminate schedule.

The principal advantage of this form of construction belongs to the relatively large volume production boat builder because he can amortize the cost of making the mold for the liner over a large number of boats. The advantage is one of cost. For example, a well-engineered hull liner can be used effectively to stiffen or reinforce the hull itself, providing an alternative to using a thick solid laminate or a cored construction for the hull. Instead, the builder uses the liner to stiffen a relatively thin hull laminate—not the best approach for providing impact resistance, perhaps, but light in weight and lower in cost.

The second major cost saving from use of hull liners is a reduction in the labor content of the boat. Building the interior of a boat is the most labor intensive part of the production process—even with the use of hull and deck liners. Despite the upfront cost of producing the molds for hull and deck liners, the ability to pop the liners out of the molds with a minimum of labor and relatively low cost ingredients over a fairly long production run means that it costs less to build the boats this way. And in the highly competitive boat market, that cost savings gets passed on to the buyer.

In larger boats (over 40 feet), hull liners may be molded in two or three sections. One may include the galley and main salon, another, the after cabin and head, and a third the forward cabin and head. In boats up to the 40-foot range, it is common for a hull liner to include the cabin sole, quarter berths, galley, setees, cabinets, shelves, head, and vee berth (Photos 9 and 10). Oversimplifying only a little bit, once the liner comes out of the mold the builder merely attaches a few pieces of teak

9. Both hull liner and deck liner are used to provide the interior finish and furnishings.

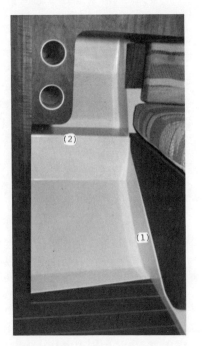

10. The ribs (1) and stringers (2) molded into this section of hull liner are engineered to provide structural support to a relatively thin hull laminate.

and fabric trim to break up the "plastic" look, runs wiring and any plumbing lines behind, around, or under the liner, and then places the entire structure in the hull, fixing it in place with polyester putty wherever the hull and liner surfaces join. Usually, grooves or ridges are molded into the liners wherever bulkheads belong. The precut pieces of veneered plywood are positioned in these grooves and fastened in place with machine screws. Drawers and cabinet or stateroom doors are installed. Galley and head fixtures are installed. And various engine, plumbing, and electrical installations are completed to the extent possible before the deck and deck liner go on. What has been omitted is the time-consuming job of building the interior furnishings piece by piece, bonding each component to the hull.

But there are trade offs for those lower costs. Because the wiring usually is run behind the hull and deck liners, it is inaccessible for repair. Over a period of time, the wiring may also be vulnerable to chafe between the hull or deck and the liner. Deck fittings often are mounted on the deck and then permanently hidden beneath the deck liner. Access to through-hull fittings may be severely limited; access to other areas beneath the cabins may not even be possible without cutting holes in the sole. Leaks are almost impossible to trace back to their source. Moreover, repairing a leak at a deck fitting, or replacing a fitting, may not be possible without cutting through the deck liner.

Frequently, builders rely on the hull liner to stiffen hull sections; the hull itself can be made comparatively thin to reduce the weight. While this hull laminate may be adequate from the perspective of tensile and flexural strength required to accommodate normal wear and tear, it may be less than adequate from the perspective of impact strength. In the event of damage from impact—striking a rock or reef, or hitting a semisubmerged log, railroad tie, shipping container, or other large piece of floating debris—the relatively thin hull laminate may be more vulnerable to puncture or holing; in addition, the damaged area of the hull may be inaccessible behind the liner. In good weather, it may well be possible to effect emergency repairs to stem the flood from outside the boat; in bad weather—the most likely time for such a problem—it could cost someone's life to try to make repairs from outside the hull.

For all of these reasons, we believe most boats made using full hull liners are best suited for dockside living or for sailing in protected or semiprotected waters. Some boats made with full hull and deck liners, however, are so well done in every other respect that they are suitable for limited coastal or near offshore cruises despite their FRP hull liners—as long as you are able to pick the weather conditions for your cruise with a reasonable probability that the forecast will hold. These boats would not be our choice, however, for extended coastal or offshore sailing.

Partial hull liners: A partial hull liner can represent a good compromise between the need to economize on labor and the need to produce boats suited for more challenging sailing than most boats made with full liners are suited for. The partial liner may consists of an FRP pan (a molded cabin sole having bumps, dimples, and grooves designed in to locate the interior furnishings) and a few parts of the interior furnishings molded in—for example, settee bases and a frame for the forward vee berth.

Alternatively, it may be a complete head or galley assembly that can be placed into the hull, with the rest of the interior built in around it. In the latter instance, the partial liner enables the builder to reduce the labor content of his interior by molding discreet portions of the interior assembly. In the former instance, the expanded pan can be used as a framework for constructing essentially the entire interior assembly outside of the boat where it is much easier to work (Photo 11). The veneer is removed from plywood wherever FRP bonds will be made to the hull; the wiring is run throughout; plumbing is completed to the extent possible; and the entire pan assembly is then placed carefully into the hull. With the assembly properly positioned, the edges of bulkheads, furnishings and the pan are bonded to the hull, using fillet bonds wherever appropriate.

In contrast to the traditional approach for installing interiors piece by piece, this use of partial liners (and full liners) lets builders construct the interiors of their boats in a parallel production line so that all three major components (hull, deck,

11. Bulkheads, interior furnishings, wiring, and some plumbing can be preassembled on the FRP pan and then placed into the hull as a unit. Veneer has been removed from the bulkheads and furnishings where they will be taped to the hull.

interior) are completed at about the same time. In addition, the carpentry can be done in the shop with maximum efficiency, without the added difficulty of assembling the elements in the boat hull. This not only shortens the elapsed time required to build a boat, it reduces the labor content of the assembly process, thereby reducing the overall boat cost. At the same time, the interior components can be taped securely to the hull and good access can be provided to wiring, plumbing, deck fittings, and through-hulls, thus working around many of the shortcomings of full liners. We still have some concern about access to the bilges through a partial liner; the key is having a sufficient number of access hatches in the pan to reach most areas of the bilge.

CONSTRUCTION DETAILS

It is generally difficult if not impossible to assess the quality of a hull or deck laminate in detail unless you are at the factory to watch your boat be constructed. As we've noted several times, however, it is possible to measure the overall quality of a boat and its suitability for the sailing you want to do by looking at details of construction that can be seen on a finished boat and using those details as a yardstick for measuring the boat. One critical detail already discussed is the system used to provide an interior framework to support the hull and deck. Others include the hull-to-deck joint, through-hulls, windows and hatches, deck fittings, steering, rigging, plumbing, wiring, and the engine installation.

Hull-to-Deck Joints

Few details of boat construction are more critical than the hull-to-deck joint. The fundamental fact is that no matter how strongly the hull and deck are built, the boat will only be as strong as the joint formed where the hull and deck come together. If the boat never leaves the dock, the strength of this joint probably is not very important. Once away from the dock, however, it's a different ballgame.

The difficulty boat buyers face in discussing hull-to-deck joints with builders is that some builders develop their own tests and cite the expertise of their engineers, or testify from their own experience to show that their way of doing the hull-to-deck joint is "as good as" or "better than" more conventional, time-proven methods of making the joint. This either pits you against the engineer who designed the joint or puts your opinions next to those of someone who claims far greater experience than you have. The only defense is to stand closely by the accumulated wisdom of sailors who have no vested interest in your decision, and that's what follows.

Hull-to-deck joint configurations: Most boats are constructed using one of four basic configurations for the hull-to-deck joint (Fig. 8):

(1) The deck is laid upon an inward turning horizontal hull flange which runs along the sheer line of the hull and is then fastened in place.

(2) The deck is laid upon an outward turning horizontal hull flange.

(3) The deck has a short descending flange and rests upon the edge of the hull laminate much like the lid of a shoe box sits atop the bottom part of the box.

(4) The deck has a descending flange that is several inches or more deep, the bottom inch or two of which overlaps the edge of the hull, forming a vertical joint. This design usually is adopted to provide more headroom in the cabin, perhaps to make use of an existing hull mold or simply to provide the appearance of a lower hull profile.

In general, both the shoe-box and vertical joints are suitable principally for dockside living or for sailing in good weather on protected or semiprotected waters. These joint configurations are vulnerable to impact because the joint is on the outer surface of the hull. More importantly, the forces on the joint that are generated as the hull goes through even one- or two-foot waves are shear forces, stressing the joint in its weakest dimension. If all other things were equal, the shoe-box configuration would be preferable to the vertical joint from a structural viewpoint because some of the vertical forces are absorbed by the

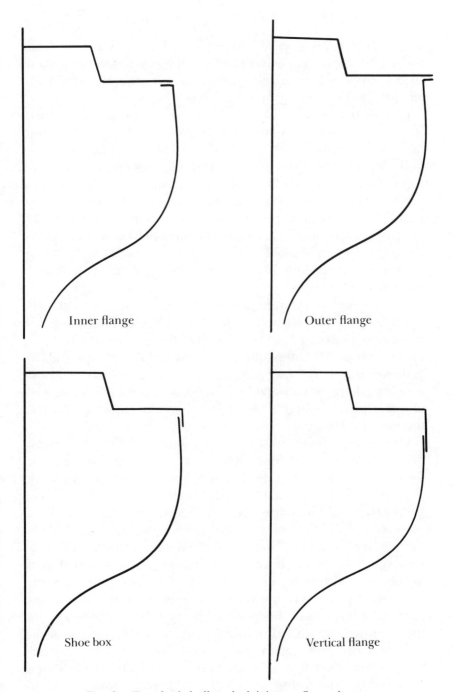

Fig. 8. Four basic hull-to-deck joint configurations.

deck where it rests on the edge of the hull. The fact of boat-buying life, however, is that all other things are never equal and such elements as interior space may override the structural advantage—as long as the boat will be kept in suitable waters.

Our own preference for offshore or coastal sailing is for a hull-to-deck joint made using an *internal* hull flange. Although this configuration is the most costly method for making a hull-to-deck joint, it provides maximum strength because (1) the major forces on the joint are absorbed by the laminate rather than by the fasteners, and (2) the joint is protected by being inside the hull. The increased expense of this joint stems from the molding process. As noted earlier, the builder must put strips along the edge of the hull mold to form a separate mold for the flange before laying up the hull, and then remove those strips before the hull can be pulled.

The *external* hull flange provides a lower cost way to achieve similar strength in the hull-to-deck joint. It is lower cost because it does not require any extra work; the flange is an integral part of the hull mold. Unless there is a sturdy rub rail either above or below the flange (as opposed to covering the flange), however, the joint is vulnerable to damage because it protrudes beyond the hull lines (Photo 12). It is this vulnerability that makes the external flange our second choice.

Hull-to-deck joint fasteners: Builders of FRP boats today generally use one of four basic means to fasten the hull and deck of their boats together, or some combination of the four: pop rivets, screws, bolts, or an FRP bond.

Pop rivets: While pop rivets may be fine for many purposes, they are not the method we would choose to make the hull-to-deck joint of a sailboat because of the dynamic loading the joint undergoes. If the rivets are not absolutely tight, they will work in the joint, enlarging the hole in the laminate until they can pull free. Even where the rivets are tight when the boat is first constructed, they may be loosened in time. If, as in many boats, the rivet goes into an aluminum strip along the outside of the joint (a plastic rubbing strip is generally fit onto the aluminum to cover the rivet heads), corrosion between the stainless steel rivet and the aluminum is likely. In all three circumstances, the rivet may pop loose under stress. For these reasons, we believe

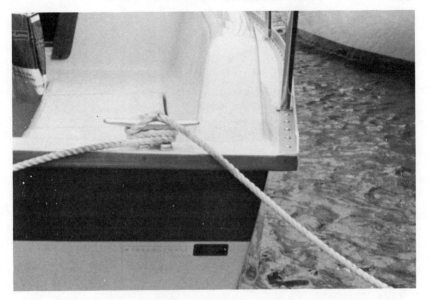

12. The external hull flange may be extremely vulnerable, overhanging the hull as in this boat by nearly two inches.

boats whose hull-to-deck joints are made with pop rivets should be used only for dockside living or for sailing in favorable weather conditions in protected or semiprotected waters. Pop rivet joints are particularly vulnerable in boats whose standing rigging is anchored to the deck.

Self-tapping screws: By themselves, self-tapping screws are not suitable for making a hull-to-deck joint and we would be surprised to find any builder using them for this critical purpose on any sailboat. The effective strength of these fasteners is only as great as the bite of the screw threads into the laminate. Holes drilled a millimeter too big to make it "easier" to insert the screws, or because the correct bit broke and this was the best available replacement, or because the power drill put the screw in too fast and "stripped" the threads are only three of many reasons why a self-tapper can fail. However, self-tappers are used well by a number of builders in conjunction with bolts or an FRP bond to make the joint. In general, the screws are used to fasten the deck in place until the bolts or FRP bond can be added later.

Bolts (machine screws): We believe that machine screws make the best hull-to-deck joints. They are most likely to be used in conjunction with an internal or external horizontal hull flange. They also are often used to fasten the toe rail to the boat at the same time, serving double duty. Most often, the bolts are used with flat washers and aircraft nuts, or flat washers, lock washers, and nuts. From time to time, builders have done away with the washers and nuts by threading the machine screws into an aluminum strip which runs the length of the joint. The first approach is preferable because the bolts, nuts, and washers are all stainless steel. The second approach works fine as long as there is no leakage; however, any leak around the joint may lead to corrosion of the aluminum around the stainless steel machine screws, and potential failure.

FRP bond: Of the four methods for fastening hull-to-deck joints, the FRP bond—a system in which the hull and deck are "fiberglassed" together—is the most difficult to deal with. While the FRP bond may offer advantages for preventing leaks, it has four problems associated with it as far as coastal and offshore cruising are concerned *unless the joint is also bolted* (Photo 13).

• The FRP bond for the hull-to-deck joint is a secondary bond. Therefore, to effect a strong bond, the entire area to receive the FRP laminate must be thoroughly scuff sanded and carefully cleaned before the laminate is applied. On the exterior hull and deck surface, the gel coat must be ground away. On either the interior or exterior surface, the bonding laminates should extend a minimum of four inches onto both the hull and deck surfaces.

• If the bond is made on the outside of the hull and deck, it is vulnerable and must be protected by a good rub rail.

• If the bond is made on the inside of the hull and deck, either it must be made before most of the interior furnishings (which block access to the joint to apply the laminate) or the workers must pay exceptional attention to detail so that the bond is made correctly along the entire length of the joint. One does not have to try to look at the hull-to-deck joint from the interior of many boats to recognize the difficulty in making such a bond, and making it well. We are concerned that inadequately supervised

13. This cutaway shows how an FRP joint can be made using a cored construction—in this case, Airex foam. Note (1) that the corner has been rounded with a polyester putty or sealant to allow a smooth radius to the FRP bond and (2) that the finished joint has been bolted to ensure that shear loads do not break the secondary bond between the cored laminate and the FRP taping used to make the joint.

workers will take shortcuts in making such a bond, which is unusually difficult.
• If the bond is made on the inside of the hull and deck, it should be put in place before deck hardware is installed. Otherwise the backing plates and bolt ends get in the way of the bonding laminate. If the bond goes in first, installing deck hardware takes more time and is, therefore, more costly since it's much more difficult to install deck hardware after the deck is on the boat than when it is sitting on sawhorses out on the shop floor. If the deck hardware goes on first, the bonding laminate must be cut to fit around the backing plates and fasteners or it will just cover them up, making future maintenance on those fittings difficult.
• Depending upon the configuration of the joint, many of the major stresses on the hull-to-deck joint may be shear stresses

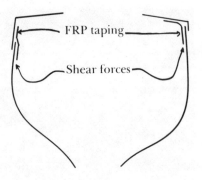

Fig. 9. The drawing illustrates FRP bonding of two different kinds of hull-to-deck joints. In both instances, shear forces created as the boat works will tend to break the adhesive bond between the hull or deck and the FRP taping.

(Fig. 9) and, thereby, may attack the weakest aspect of the FRP bond—the adhesive strength of the resin between the hull or deck and the bonding laminate.

For all of these reasons, while the FRP bond is certainly suitable for dockside living, for sailing in protected and semi-protected waters, and—if done well—for limited coastal cruising in carefully chosen weather, we would want to see it used in conjunction with through-bolts if the boat were to be used regularly for coastal or offshore sailing.

Hull-to-deck joint sealants: While safety of the hull-to-deck joint must be of paramount concern, the ability of the joint to keep rain and sea water outside the boat must run a close second. That watertightness is, more often than not, a function of how the joint is sealed. A variety of sealants exist for builders to choose from, but only a very few are well-suited to the job. Oil-based sealants, for example, are not at all suited for this use (though we have seen $80,000 boats in which they were used); they will dry out in time, crack, and leak. Their only virtue is low cost to the builder. Our own feeling is that builders should use a polyurethane adhesive and sealant like 5200 adhesive/sealant for the hull-to-deck joint. Moreover, they should use an excess, so that you can see on the inside of the joint where some of the sealant has been squeezed from the joint. A good polyurethane sealant such as 5200 will remain flexible almost indefinitely; it will adhere tenaciously to both the hull and deck surfaces; and it is strong in its own right.

As with anything else, however, the sealant must be applied correctly if it is to perform as advertised. Properly done, the mating surfaces of the hull and deck are cleaned thoroughly, eliminating all dust; a generous amount of sealant is applied; the deck is lowered onto the hull and fastened in place with self-tappers—but not screwed down tight. After the sealant has cured most of the way, the self-tappers can be tightened firmly, the bolts installed with fresh sealant around them, and the nuts drawn tight. The idea is to allow the sealant to cure as a relatively thick layer and then tighten down on the joint; if the joint is drawn tight before the sealant cures, much of the sealant will be squeezed out of the joint, possibly compromising the watertightness.

Windows and Hatches

Four questions come readily to mind when thinking about windows and hatches: Do they leak? Do they provide good ventilation? Do they provide light? And, do they provide a view of the world outside of the boat? Depending upon the use intended for your boat, the relative importance of these functions may vary. If you are considering coastal or offshore cruising, for example, hulls, decks, and hull-to-deck joints constructed to withstand icebergs are of little value if the first solid water on deck will knock out a window or pop the Lexan from a hatch. By the same token, the small, sturdy ports and windows best suited for offshore sailing have little to offer for dockside living—even on weekends.

It is relatively easy to recognize the virtues of large windows and hatches for dockside living. For those who have not been exposed to the force of waves in a major storm, however, it may be more difficult to appreciate the need for strength in window installations. Excerpts from accounts in *Sail* magazine of tragedy and near tragedy involving two cruising yachts perhaps make the point. The first recounts the experience of a 42-foot cruising yawl on a delivery trip to Florida. The boat was just a few miles off the coast of Maryland in a severe storm when the helmsman "looked astern and saw a huge wave collapse on the top of the mizzenmast. He shouted a warning and was thrown into the air and over the side to the end of his bowstring-tight tether. The boat rolled her mast under, perhaps as far as 170 degrees, and came up with the mizzenmast in pieces but the main spar intact. . . . The plywood hatch had been smashed in, and *the plastic frames for the glass ports had given way, so the boat was half full of water and vulnerable to another wave.*"[1] The crew and boat later were towed into the safety of Ocean City, Maryland, by the Coast Guard.

The second account details the loss in the same storm at the edge of the Gulf Stream off the Carolina coast of a heavy displacement, 58-foot ketch designed for comfortable world cruising. The main cabin had three large windows port and starboard, with provision for sturdy storm shutters for use in heavy weather. On this trip, storm shutters were installed to starboard, the expected windward side. After hours spent struggling against the weather, the helm was lashed, the boat left to lie ahull in the breaking seas, and the crew made themselves secure below. As one of the crew watched through the unshuttered leeward windows, the boat "rolled and the leeward rail went under, and then the windows themselves went under; then the 30-ton ketch lurched farther and fell off a wave onto her lee side, and [he] saw the Atlantic Ocean pour through three broken windows."[2] Shortly after, the crew watched the ketch sink from their life raft. In the four days that followed before the life raft was sighted, two of the crew became delirious and left the life raft to drown; a third died of injuries received when thrown across the cockpit of the ketch before it had sunk. Only two survived.

Windows: The most effective windows for keeping water on the outside are fixed ports. They potentially offer the greatest strength factor, particularly if they are bolted directly to the cabin house. Moreover, fixed ports can't be left open either for rain or spray to blow inside, or for sea water to rush in if solid water comes aboard. It's a good idea, however, to begin any consideration of windows with one basic assumption: Sooner or later, all windows on a boat will leak and fixed ports are no exception. For that reason, they should be easy to remove for rebedding. That means that all fasteners used to install fixed ports should be readily accessible. It also means the windows

---

1. John Rousmaniere, "Death in the Sudden Gale," *Sail*, May 1983, pp. 88–92.
2. Ibid.

should be made of a material such as Lexan that is strong enough to survive being removed. If fixed ports are installed using a frame of aluminum or other material, not only should the frame be readily removable for rebedding, but the gasket sealing the window within the frame also should be easily replaceable.

In addition to letting water in if they are left open too long, opening ports suffer the same leakage problem found with fixed ports. As a result, they too should be easily removable without breaking. Too often, that is not the case. Many boats today are being delivered with plastic opening ports which break easily when being removed for rebedding. As a result, fixing leaks can get expensive. Recognizing that one caveat, the plastic opening ports are certainly suitable for dockside living and sailing in protected waters. They are probably also suitable for most sailing in semiprotected waters. In our view, however, plastic opening ports should not be used on a boat intended for coastal or offshore sailing. The plastic ears used to dog these ports down can be broken easily, compromising the integrity of the port. These ears are particularly vulnerable if the window is being closed and made tight in a hurry, when you are more likely to be thinking about getting all ports closed before the rain comes blowing inside than about how hard you're cranking down on the dogs. In addition, we are concerned that these windows would not withstand the direct impact from a breaking wave, from the boat falling from a wave onto its side, or from being rolled. For these reasons, we want to see boats intended for use in coastal or offshore waters fitted with opening ports made of bronze or sturdy cast-aluminum frames, or with sturdy fixed ports.

When ventilation is considered, opening ports have one major advantage: In a marina, they will permit cross ventilation if any air is moving. Under way or at anchor, when ventilation tends to be fore 'n aft, hatches are most effective—if they can be kept open. If the forward hatch must be kept closed, opening ports may be helpful in promoting air flow. When considering the amount of light they allow into a boat and the field of view they allow from inside out, fixed ports have the advantage: They can be made larger and, for a given size, stronger than the opening ports normally used in boats we are considering. Once again, however, it's necessary to weigh the intended use of the boat against window size. As we have noted several times earlier, large windows are suitable only for dockside living and for sailing in protected or semiprotected waters *unless there is provision for installing strong storm shutters to protect the large windows from the impact of solid water.* It is also necessary to weigh the method used to install the fixed ports against the use intended.

Hatches: With one important exception, the configuration of hatches is relatively unimportant from the perspective of safety when considering a boat for dockside living or sailing in protected waters. Convenience of access to hatches and ventilation are likely to be major considerations in these boats. The exception, which will be discussed in detail below, is the availability of an "escape hatch" on the foredeck.

Hatches become progressively more important, however, as you consider a boat for use in semiprotected waters, where quite rough water may be encountered even though help is relatively close at hand, or in coastal or offshore waters, where there may be no escape from heavy weather conditions until the conditions themselves ease. In looking at boats for semiprotected, coastal, and offshore cruising, we are concerned principally about three different kinds of hatches: the companionway hatch, hatches in the cockpit or afterdeck which open to lockers or to the lazarette, and cabin-top or deck hatches opening into the cabin area. The temptation in looking at a companionway hatch is to think about ease of access. A nice wide companionway with a simple step down from the cockpit sole is certainly the easiest pathway into the cabin. Easiest is not necessarily best for these waters, however. The companionway hatch should begin about at cockpit seat level so that you must step over the "bridge deck" to reach the companionway ladder. This bridge deck serves as a dam to help keep water out of the cabin if a wave invades the cockpit. Moreover, the companionway should only be wide enough for one person to pass through without contortions. Again, the question is, how big a hole do you want for water to come into the boat? The hatch boards used to close the companionway should be at least ¾-inch stock, and there should be some way that they can be locked in place from within the cabin so they will not fall out in a knockdown. Finally, in our view, every companionway hatch should have a sea hood—a hood into which the sliding top of the hatchway slides when it is opened. The sea hood helps keep spray and rain water from running under the sliding hatch and dripping into the cabin.

Two key features are required of cockpit locker hatches or lazarette hatches on boats intended for use in semiprotected, coastal, or offshore waters: (1) They need a latching system to ensure that they will stay closed even if the boat is knocked down and the contents of the locker fall against the hatch cover. Obviously, a simple friction latch will not suffice. One alternative is a hasp and eye snap on the cockpit locker and lazarette hatches; the eye snap lets us get into the lockers easily if necessary, but provides a strong, positive latching system in the event of a knockdown. Moreover, in port, the eye snap can easily be replaced by a padlock. (2) Cockpit or aft-deck lazarette hatches should be water resistant; preferably, they should be reasonably watertight. Water resistance is particularly important if the hatch extends very far down toward the cockpit sole (Photo 14). Otherwise, a flooded cockpit will drain into the lockers, not out

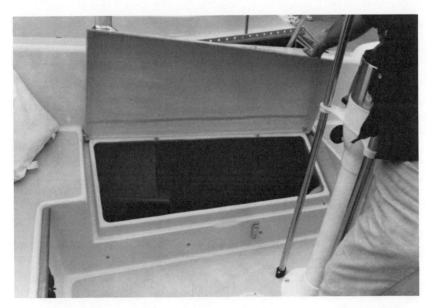

14. The low edge of this cockpit locker suggests a boat suited for protected or, possibly, semiprotected waters.

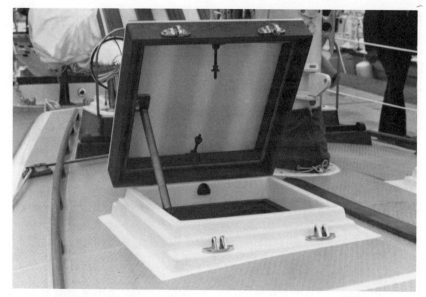

15. An excellent hatch noteworthy for its two step coaming, which allows a snug-fitting hatch cover to provide maximum protection against water leakage.

the cockpit drains. In many boats we've seen, the boat's engine and batteries would be flooded in those circumstances. In others, the water would flood into the cabin through the locker.

Deck and cabin top hatches have three principal requirements: *One requirement which applies to all boats with an enclosed cabin is that at least one deck or cabin-top hatch be big enough for the largest crew member and reachable by the smallest crew member for use as an escape hatch in case of a fire blocking access to the companionway hatch.* The second requirement is that all such hatches be watertight. The third requirement is that the hatches be sturdy enough to take whatever abuse comes their way (Photo 15). With most builders using aluminum frame hatches with Lexan covers, we recommend insisting upon well-known, brand-name hatches. Bomar and Goiot are among the best known. Some builders have hatches which look much like the brand-name products but in fact are only lower-cost lookalikes. Although it is difficult if not impossible for us as boat buyers to measure the quality of those hatches, the question is worth pursuing. We have seen a 50-foot ketch built by one of the more reputable Florida boat builders arrive at a Hilton Head, South Carolina, marina with the Lexan panel blown out of the aluminum frame of one of the foredeck hatches. It may be coincidence, but that hatch was one of the unbranded variety.

Through-Hull Fittings

Perhaps more than any single part of a cruising sailboat—or large power boat, for that matter—the central question surrounding discussion of through-hull fittings is summarized in the phrase, "margin of safety." All of the issues revolve around that question: What margin of safety do you want for your boat? It comes down to the simple fact that if there is a failure in the through-hull fitting system, the boat can sink—even sitting at a dock. The issues involved include: the kind of through-hull system, accessibility, shut-off valves *versus* no shut-offs, sea cocks *versus* gate valves, and metal *versus* plastic.

Accessibility: No matter how you plan to use your boat, we believe all through-hull fittings should be readily accessible. Unfortunately, too many builders—or designers, we're not sure which—disagree. The result is that too many boats are constructed in which it is difficult at best and sometimes even virtually impossible to get to all through-hull fittings from *inside* the boat. Part of the problem stems from the use of hull liners. In construction, the builder may install and run hose from a through-hull fitting before the liner is installed. In some cases, the liner may cover up the through-hull completely; in others, the through-hull is positioned so that no ordinary human being can reach it once the boat's interior has been completed. Or it is located at the bottom of a cockpit locker which, when the boat is being used, will be kept filled with gear—all of which must be removed to get at the through-hull. It may even be necessary to climb down into the locker to reach the through-hull. Under these circumstances, consider what happens when you wake up at night while week-ending on your boat in the marina, or at sea in stormy weather, or anywhere between those two extremes and find water ankle-deep in the cabin. The first thing you need to do is check your through-hull fittings to be certain a hose hasn't come loose. And it is pitch black. We can guarantee you will have a problem.

The alternative is to select a boat in which you have easy access to all through-hulls. Moreover, if the boat is planned for sailing in semiprotected, coastal, or offshore waters, all through-hulls should be accessible from inside the cabin. One would prefer not to open a large cockpit locker in the midst of a storm to reach a through-hull. The accessibility imperative applies also to through-hulls located above the waterline. It is quite possible, for example, to flood a boat by having water back-siphon through a bilge pump when the bilge pump through-hull is buried in the water as the boat heels over. All that's required is a piece of debris blocking the valve in the bilge pump that is supposed to prevent back-siphoning.

Shut-offs: Most if not all builders agree that every through-hull fitting below the waterline should have some kind of shut-off valve. Many—if not most—builders, however, appear to believe that shut-offs are unnecessary *above the waterline* (AWL). For dockside living, one can make a reasonable case for omitting the expense of AWL shut-offs. Once the boat leaves the dock, it becomes more difficult to justify omitting the AWL shut-offs. Many through-hulls that are above the waterline when a boat is at rest are quickly put below the water's surface by the hull wave that forms as the boat gains speed through the water. Moreover, as the wind fills the sails and the boat begins to heel, more through-hulls that are located above the waterline may be put underwater. Obviously, the issue should not be whether the through-hull is above or below the waterline, but whether it will be above or below the water level with the boat in normal use. For sailing in semiprotected, coastal, or offshore waters, we'd insist that all through-hulls have shut-offs. Moreover, we recommend keeping all shut-offs closed except when they are in use.

Sea cocks *vs.* gate valves: We hesitate to lend credibility to the use of gate valves by acknowledging that this is even an issue. But it is an issue because there is a significant difference in cost between the two. At this writing, the full retail price of a ¾-inch bronze sea cock is between $50 and $60; the same size gate valve costs about $7.50. From the perspective of safety, however, there's no contest between the two. Gate valves have the unpleasant habit of freezing in position from corrosion between the dissimilar metals in their construction. As a result, gate valves have proven notoriously unreliable over the years; on the other hand, sea cocks have demonstrated notable reliability.

Metal *vs.* plastic: In the past 10 years, plastic through-hulls and plastic shut-off valves have found increasing use on boats. The reason is cost. Depending upon size, the cost of a plastic sea cock is from one-fifth to one-third the cost of a bronze sea cock. The cost of a plastic through-hull fitting may be as little as one-tenth the cost of the same size bronze through-hull.

We have a bias against plastic in these applications. We have experience with plastic in sea cocks which justifies our bias: The plastic handles break. If the handles break when water is flooding into the boat through a broken hose, you've got a problem. More likely, it breaks in normal use, but doesn't get replaced because the valve is left in the open position. In that situation, you might as well not have the valve if an emergency arises.

We don't have the same kind of experience to justify our bias against plastic through-hulls. We do have that bias, however, and recommend use of bronze through-hull fittings below the waterline on all boats, and both above and below the waterline on boats for coastal or offshore sailing. All bronze through-hull fittings located below the waterline should be connected to the boat's external ground to prevent electrolytic damage. Plastic through-hulls may be suitable for use *above the waterline* on boats intended for dockside living or sailing in protected or semi-protected waters where help or shore or both are nearby (Photo 16). In any case, they should be kept well clear of any possible heat source—including any possible shipboard fire. One boat owner has reported, for example, that his boat nearly sank when a plastic through-hull melted. The problem arose when the water pump on his engine failed. Before he was alerted to the engine overheating, the exhaust system became so hot that the nearby plastic through-hull was melted and the flooding began.

FRP through-hull systems: Most through-hulls are mushroom-shaped fittings made of bronze or plastic. Some builders, however, have developed a system of fabricating through-hulls by fitting a fiberglass tube into a hole cut in the bottom of the

16. This anchor well drain is located where it will be in the way of people's feet if the vee berth is used for sleeping. The plastic hose cuffs are fastened to plastic fittings from the well and through the hull with only single hose clamps. We question the suitability of such an arrangement even for use in protected waters.

boat. In general, this system is used for through-hulls above the waterline. In practice, however, the opening is often so close to the waterline that it is underwater most of the time while the boat is under way. Typical applications include cockpit drains, the rudder post through-hull, and drains for foredeck anchor wells (Photo 17). The problem is that these tubes either crack,

17. An FRP tube is used to bring the rudder shaft through the bottom of this boat. We have known of tubes used in this manner to break away from the hull and cause serious leaking.

break or break loose, and leak. One 30-foot boat we know had the tube used as the through-hull for its rudder post break loose from the hull as it was sailing up Delaware Bay. When the source of the leak was discovered, all of the crew but the helmsman crowded onto the bow to lift the stern out of the water until the boat made its way to a marina. The same couple whose boat suffered an electrical fire in mid-Atlantic from wires chafing between the hull and hull liner (Chapter 5) had another experience which illustrates the weakness of this system for making through-hulls.

> ... I could now clearly see the source of our mysterious flooding. [Our boat] has a recessed anchor locker in its foredeck, and this locker is drained by two strong, three-inch scupper pipes, which exit from both lower after corners of the molded box. Lying on my belly on a wet sail bag, I reached up to touch the fiber glass tube of the portside scupper. Cold sea water ran down my fingers and palm and under the cuff of my sweatshirt. The drain pipe was neatly broken, completely fractured about two inches from the point at which it joined the hull, maybe a foot and a half above the waterline. Heeled as we were on this tack, the bow wave sloshed back and up and into the drain pipe, then right into the boat through the fracture.... The terrible weather we'd ridden out had twisted the hull badly enough to not only short out that lighting-wire bridle under the hull liners, but also to fracture this scupper pipe, way up there.[3]

Because of their inherent weakness, this or any similar system of making through-hulls with fiber glass tubes is unacceptable by our standards for any but dockside living and sailing in protected waters.

Standing Rigging

Most of us are unlikely to develop the engineering expertise to determine whether a builder is using adequately sized wire or rod rigging to support the sail plan of any particular boat. If the boat was designed by an independent yacht designer, one

---

3. *First Crossing: The Personal Log of a Transatlantic Adventure*, p. 148.

can always telephone the designer to find out what rigging he or she specified and then compare the answer to the builder's specifications. If, on the other hand, the design was done by the builder's in-house designer, that query isn't likely to get you very far. As a result, the best you can do is compare the size of the rigging on one boat with that on others, keeping in mind that whether any particular set of wire or rods is adequate depends in part on the use intended for the boat.

Some racing boats, for example, are built with minimum wire or rod size to help reduce both windage and weight aloft. That may be fine on a boat crewed with experienced sailors who recognize the risk of losing the rig. That's not so fine, however, on a boat intended for family sailing. A sudden squall or thunderstorm might catch a less experienced crew napping. That's also not so fine on a boat intended for coastal or offshore cruising, where losing a rig is a serious problem at best. So we compare rigging sizes, making the comparison between boats of similar displacement and sail plans, and working under the general assumption that the rigging wire should be sized to the use intended for the boat. This means that boats intended for coastal or offshore sailing will need heavier wire than those intended for use in semiprotected waters, and that boats for semiprotected waters need heavier wire than those intended for use in protected waters or dockside living. Interestingly, the differences in wire size from one boat to another are notable.

There is more that we can look at as well. Those lengths of wire or rod rigging are only one part of a larger system making up the standing rigging. That system includes tangs, turnbuckles, spreaders, chain plate attachments, stemhead attachments for headstays and stern fittings for backstays. In a design featuring a bowsprit or pulpit, the rigging system also includes the bobstay, the fitting used to attach the bobstay to the hull (Photo 18), and the whisker stays and attachments. If any of these elements is less than adequate, the entire rigging system can fail just as easily as if a stay or shroud breaks.

Chain plates: In more traditional crusing designs, chain plates are tied into the hull, usually bolted to the hull with backing plates on the inside. In more modern designs, the chain plates have been moved inboard, piercing the deck and tied into some

18. The sturdy fitting bolted to the bow provides a good anchor to this bobstay.

form of interior structure designed to withstand the various forces on the shrouds. In both arrangements, there are time-tested systems which, if well executed, are suited for any kind of sailing. In general, the chain plates in these more conventional systems are readily available for inspection and maintenance. By way of contrast, there are also variations that have been developed either to reduce construction costs or to accommodate changed interior arrangements; some of these variants, however, make it difficult if not impossible to inspect and maintain the chain plate attachments. For this and other reasons, the suitability of some variants for some sailing is open to question.

External chain plates in their most common form are recognized by metal straps bolted to the outside of the hull. They are usually backed up by a main structural bulkhead supporting the sides of the hull against the compressive forces of the shrouds—forces which tend to squeeze the hull and deck. Some boat builders add one or more extra laminates to the hull in this area because of the high stresses carried by the chain plates and transferred to the hull. More often than not, one can look at external chain plates from outside and then from within and have a good sense of what kind of waters they are suited for. If they are made of heavy stainless steel and fastened with five or six half-inch or larger bolts, a sturdy backing plate, lock washers and nuts, the chain plates probably are suited for coastal or offshore sailing (though the rest of the boat may not be). Less sturdy arrangements will be best suited for semiprotected or even protected waters. At the least, the chain plates should be sturdy enough that they do not flex under load. They also should be angled so that they lead straight up into the shrouds. In all instances, the chain plates should be attached to the hull and not to a descending deck flange or to the cabin sides. Such an arrangement puts the rig's forces on the hull-to-deck joint.

A variation on the external chain plate theme involves embedding the chain plates within the hull laminate. For example, (Fig. 10) halfway through the laminating process, a chain plate with two or more cross pieces may be fiberglassed into the lay-up. In theory, with this approach the chain plates can be made extremely strong. Long strands of roving are draped over the crossbars of the chain plate and streamed down the side of the hull laminate, and then covered along with the chain plate by the next layer of reinforcing fabric.. Unfortunately, the system is flawed. And, the problem is water.

Chain plates work when a boat is under sail. Moreover, some of the forces on the chain plate tend to separate the laminates. As a result, sooner or later, that working of the chain plates will let water get into the hull laminate. In some boats we know of, the chain plates have started rusting inside the laminate. In others, the water in the laminate has frozen in the winter. In both circumstances, the expansion caused by rusting or freezing has aggravated the leakage problem and eventually, expensive repairs were required. In one series of one-design racing boats, whose chain plates were embedded in the hull laminate, the builder wound up replacing the chain plates in a number of the boats.

If the boat is used principally for dockside living, or for sailing only in protected or semiprotected waters, this system may be adequate with good maintenance of the sealant where the chain plates come through the deck. If the boat is intended for

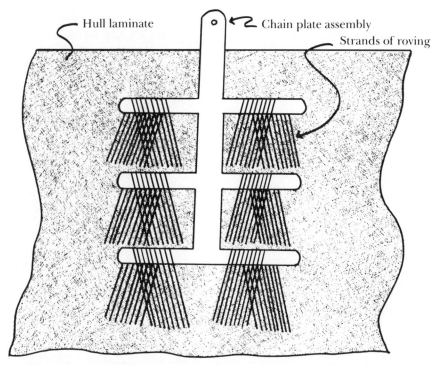

Fig. 10. Chain plates embedded in the hull laminate usually have two or more crossbars. Strands of fiber glass roving are wet out with resin and draped over the crossbars to help anchor the chain plate. Subsequent layers of fiber glass fabric bond the entire assembly in the laminate.

coastal or offshore sailing, however, or even for frequent sailing in strong winds in semiprotected waters, we prefer more conventional chain plate systems.

Inboard chain plates are far and away the more common chain plate system used today, most likely the stepchild of the relatively beamy designs spawned by the IOR rules. While inboard chain plates offer the advantage of tighter sheeting angles for better windward performance, they have two inherent disadvantages: (1) They are toe breakers on deck and, therefore, one of the best arguments around for requiring all hands to wear shoes on deck; and (2) eventually they will leak, though that too is a problem which can be handled with a bit of preventive maintenance (Photo 19).

19. These chain plates are out of line with the shrouds. In addition to concern about the misalignment of rigging forces, this should raise questions about other places where attention to important detail may be less than it should be.

Traditionally, inboard chain plates for the upper shrouds are bolted to the boat's main bulkhead. Chain plates for the lowers are tied into secondary bulkheads or knees. Done well, this is an excellent system. However, both the bulkheads and the knees should be bonded not only to the hull, but also to the deck and cabin house. In boats over about 25 feet and up to about 42 to 45 feet, the main bulkhead should be of ¾-inch marine grade plywood. In larger boats, the main bulkhead should be 1½ inches thick. The secondary bulkheads or knees should be the same thickness as the main bulkhead. In addition, a good practice used by a few builders involves laying up an FRP strip about ⅛-inch thick on each side of the bulkhead where the chain plates and backing plates will be bolted (Fig. 11). The FRP laminate is not needed for added strength in the conventional sense. Rather, it helps prevent even the possibility of the bolts working in the relatively soft plywood and making their holes larger. It can

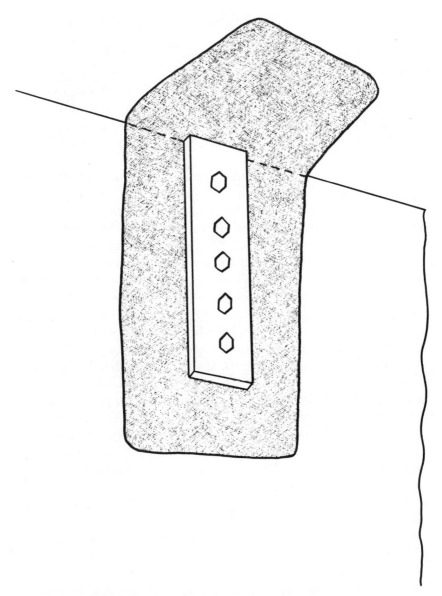

FIG. 11. An FRP pad is sometimes used to reinforce the bulkhead chain plates are bolted to. As illustrated, the FRP pad can be extended to the underside of the deck, further strengthening the installation.

also be used to help bond the bulkhead to the underside of the deck.

More recently, a number of builders have begun replacing bulkheads as anchors for chain plates with a tie-rod system which connects the chain plates to a frame member set down beneath the cabin furnishings. In some instances, the tie rod is connected to the floor pan itself. This arrangement is satisfactory for much routine sailing. However, we have four principal concerns about these systems:

• The tie rods only transmit the vertical forces to an anchoring system. They do not resist the compressive forces attempting to squeeze the sides of the boat toward the centerline. Unless some kind of rigid beam is molded into the deck or deck liner running uninterrupted from port chain plates to starboard chain plates, these forces will tend to flex the deck, the hull-to-deck joint, and the hull. Some builders may use a tie rod connecting the keel and deck to keep the deck from flexing; the effectiveness of that solution to the problem, however, depends upon the overall stiffness of the deck and cabin house structure. That tie rod can merely become another focal point for deck flexing.

• In a number of boats, all of the shrouds on each side are gathered into one chain plate which is anchored below decks using a single tie rod (Photos 20 and 21). In such a rig, all of the forces on shrouds are held by a few threads on the tie rod into a nut or piece of steel plate. Although an engineer may say those threads are strong enough—and they probably are for most of sailing—we would be uncomfortable with such a rig in severe weather offshore.

• We know of instances in which the nuts used to effect the bottom connection on a tie-rod system have come loose. In a trial sail of one 39-foot boat, for example, the chain plate was almost pulled through the deck before the deck's flexing revealed that there was a problem. In addition to possibly losing the rig in such a circumstance, one could well find a sizeable hole in the deck where the shroud / chain plate / tie-rod assembly pulled loose.

• In most of the boats we've seen that use the tie-rod system for anchoring shrouds, the nut or steel plate which anchors the tie

20. The single chain plate does not provide the same fore-and-aft support traditionally provided by lower shrouds to keep the midsections of the mast from pumping in a seaway.

rod is inaccessible. As a result, that critical fastening cannot be checked easily.

Headstay and backstay attachments: It would seem obvious that the attachments for the headstay and backstay should be at least as sturdy as those for the shrouds. If one notes that the

21. The single chain plate shown in Photo 20 is connected to a stringer molded into the hull liner by a tie rod from the underside of the chain plate.

wire or rods used in these stays is heavier than the shroud wire, he might conclude the attachments should be even sturdier. Yet we see boats whose headstay is anchored to a small plate fastened to the deck with a few small bolts and with no backing plate, or whose split backstay is anchored in small U-bolts at the aft corners of the deck with only a small backing plate under the deck and much of the stress absorbed by the hull-to-deck joint (Photo 22). Such boats are best suited for dockside living or sailing in protected waters in good weather. For heavy weather use offshore or inshore, the anchor systems for the main stays should be made of sturdy materials, strongly through-bolted, tied into both the hull and deck, and accessible for maintenance. A less sturdy system can be used for fair weather sailing in semiprotected waters, but one should not be lulled by the notion of "fair weather" sailing. Weather forecasting is too unreliable and the crew's personal safety too valuable.

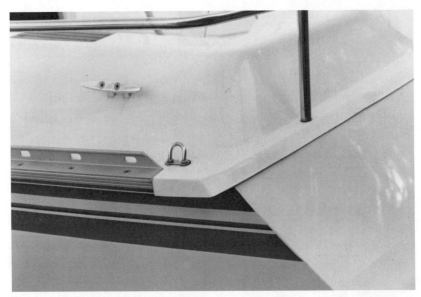

22. The split backstay of this boat's rig attaches to the U-bolt shown here.

Rudder Installations

Almost any rudder configuration is suitable for dockside living or sailing in protected or semiprotected waters in good weather. As the weather acts up, the demands on the rudder system increase, but if something in that system fails, help is relatively near. For ocean sailing—coastwide or offshore—we prefer either a traditional rudder hanging on a long keel or a rudder attached to a sturdy skeg extending the full depth of the rudder. Spade rudders, while unquestionably faster and more responsive, are best suited for racing boats or for cruising inland waters. In our view, spade rudders are too vulnerable for use offshore in a cruising boat. They are more susceptible to damage from impact, and all stress is focused on the rudder post.

In any system, a number of details must come together to provide a good rudder. There should be, for example, a stainless plate or several stainless rods welded to the rudder shaft and enclosed in the fiber glass molding for the rudder. These rods—like the steel plate—keep the rudder from rotating on the shaft. In addition, the rudder shaft should extend well up into the hull, up to deck level if possible. This shaft should be heavily braced near the top as well as where it enters the hull. The bracing is particularly important for spade rudders to resist the forces on the rudder. Even on a skeg- or keel-hung rudder, the bracing at the top of the shaft helps ensure there is structural support for the rudder at both ends as well as the middle.

To complete the rudder system, the boat needs a wheel or tiller. We question the use of hydraulic steering in boats of the size we are discussing—particularly for ocean waters. Hydraulic steering systems sometimes become sluggish and the helmsman loses touch with the rudder. When a wheel is installed, there should also be provision for emergency tiller steering. In all cases, the rudder shaft, the fitting (hopefully including a stuffing box) where the shaft comes through the hull, and any wheel steering mechanism should be readily accessible for repair and maintenance.

Deck Fittings

Aside from ensuring that the deck or cabin house is strong enough to withstand the forces on the fitting, there is no magic to installing deck fittings properly. It is simply a matter of doing the job right. There are four principal components in addition to the fitting itself: the bolts or machine screws used to fasten the fitting to the deck; the sealant used to bed the fitting; a backing plate used to spread the loads to a larger portion of the deck; and the lock nuts (aircraft nuts) or nuts and lock washers used to tighten down on the bolt. Either lock nuts or lock washers are used to prevent the nuts from working loose and falling off.

The importance of installing these fittings correctly cannot be overstated, whether we are looking at a boat for dockside living or for crossing oceans. Even at dockside, a boat is subject to strong currents and winds, possibly including hurricane-force winds.

Engine Installation

We begin with two assumptions: (1) that your boat will have a diesel engine; and (2) that you, like most of us, do not have the expertise to evaluate engine installations in any detail.

All of us, diesel experts or not, must live with the engine installation on whatever boat we purchase. As a result, we as boat buyers have a perspective which the boat builder should consider, but may not have. That's the perspective of someone who will have to carry out routine maintenance, trouble shoot the engine if it quits some miles from port, and pay the bills.

One basic question to be answered on boats for any waters is whether the engine is accessible for maintenance. The oil and fuel filters as well as the water pump(s) should be easily reached so that filters and impellers can be changed. If the builder does not provide a remote (bulkhead-mounted) fuel filter/water separator, there should be space for one to be installed—and serviced—easily. A remote oil filter also is advantageous. There should also be access to the fuel pump, injector pump, and injectors. Valve covers should be readily accessible. If the engine is equipped with a hand crank and decompression levers for hand starting, there should be adequate room for cranking (we recommend trying to turn the engine over with the crank). If you are planning coastal or offshore cruising—cruising that is likely to take you well away from your home port—you might also consider how easily the engine can be removed from the boat. If serious repairs are ever needed, they may be accomplished more readily (and presumably at lower cost) if the engine can be pulled easily and the work done in the shop rather than on the boat.

Fuel lines and the fuel tank vent are two other areas of interest. The fuel line from the fill pipe to the tank, and from the tank to the engine and back to the tank all should be protected against chafe. The vent hose also should be free from chafe. Equally important—some might say, more important—the fuel vent should be located to minimize the potential for water to enter the fuel system through the vent line. Boats with the vent fitting positioned between the shear and waterline are inviting water contamination of their fuel.

Every engine installation also should have a stuffing box on the propeller shaft. Some builders reduce costs by omitting the stuffing box, but we believe this is a corner that should not be cut. The stuffing box is the only means you have for preventing leaks around the shaft. Unless the engine is well above the waterline, even when the boat is heeled, the cooling water system should also have an antisiphon valve installed in line where the cooling water by-passes the exhaust manifold before entering the exhaust pipe. This valve is designed to prevent outside water from siphoning back into the engine. Boats intended for use in coastal or offshore waters should also have a sea cock on the exhaust line, which in turn should be looped up high under the deck. If the vacuum relief valve fails, having that sea cock closed when the engine will not be used for several hours or days may be the only thing between you and a very expensive repair. It is a lesson we have learned from the experience of a close friend: during a trip to Bermuda, he had water enter the exhaust system of his boat from the ocean, back-siphon into the exhaust manifold, and drain through the open exhaust valves into the cylinders and crankcase. It was a costly repair, but his experience has convinced us to close exhaust sea cocks routinely when in the ocean.

Wiring

While it requires a certain amount of specialized knowledge to tell whether the wiring in any boat is adequate for the job expected of it, there are a number of clues anyone can look for. For example, electric fixtures and outlets—including the circuit breaker panel—should be located where they won't get wet if a port or hatch is left open, or the hull-to-deck joint begins to leak. The wiring also should be neat and accessible for repair (Photos 23 and 24). More than one color of wire should be used to make it easier to trace circuits should the need arise, which it will. Wiring, too, must be able to flex with the boat as it works in a seaway, so it should be hung loose, not tight. Moreover, there should be slack at the circuit breaker panel so that connectors can be replaced without concern about making the wires too short. If all of the wires are the same size and thin, one can

23. Water coming through this open port in a rainfall or from spray can drain down into the light fixture and the electrical outlet.

24. This electric cable—solid wire of the type normally used in home construction rather than in boat construction—runs under the bulkhead taping and disappears behind the galley.

suspect the wiring is minimal and may overheat, causing a short. If the wires are different sizes, it probably is sized to the demands of each circuit.

Plumbing

The major problem in boat plumbing is that it is too often installed as if a boat were a house, i.e., as if a boat didn't move. One result has been the introduction of plastic water tanks and of rigid plastic piping. These may be satisfactory for some boats, but it is important to be aware of their limitations.

Water tanks: Plastic tanks are suitable for dockside living, daysailing, and cruising in protected or semiprotected waters. They are not suitable for coastal or offshore sailing. The principal advantages of plastic tanks are their light weight and low cost. Their principal disadvantages are the difficulty in cleaning them and their vulnerability to damage from chafe. The cleaning problem occurs because plastic tanks often are not fitted with clean-out plates. The chafe problem occurs because the plastic used to fabricate the tanks is softer than the fiber glass or wood they ride against. As a result, the tanks chafe against the hull and flooring, eventually developing a leak. The chafing may be the product of wave action and the working of the boat structure in a seaway when sailing coastwise or offshore or of vibration when under power. Similar chafing may result when sailing in semiprotected waters as well; the critical difference, however, is in the consequences of having your drinking water leak into the bilge: offshore you could be in a bind; inshore, fresh drinking water is at most a few hours away.

Water piping: Rigid plastic piping has been used increasingly in the past few years, in part because it is relatively inexpensive. This piping is acceptable for dockside living in a warm climate, though we would be concerned about freeze-damage in colder climes. It is also acceptable for sailing in protected and semiprotected waters. It is not acceptable for coastal or offshore sailing.

Rigid plastic piping has three principal advantages. It is neat in appearance, easy to run (before the boat's interior has been built), and it's lower in cost than the flexible hose alternatives. Its principal disadvantage for coastal or offshore use is its rigidity. The boat works in a seaway, particularly in rough water, and the water lines must be able to flex with the boat. Also, the rigid piping is difficult to repair if a problem develops because you cannot simply pull it out to fix it. Soft or flexible plastic hose accommodates the boat's working and is easier to take out for repairs, but it too is not without its limitations. Because it is softer, it needs to be protected from chafe and should be checked regularly wherever chafe is likely—even where it is unlikely.

Hose clamps: All hose connections should be made using two stainless steel hose clamps wherever space permits. Without exception, all hose connections leading to a through-hull fitting should have two hose clamps. The second clamp provides a backup against the failure of the other clamp.

7

Boat Design
A Look at Form and Function

If boat construction appeals to the rational part of our being, boat design lays claim to our hearts. We can be cold and calculating when looking at how a bulkhead is installed, but we soar on a natural high as we watch a nicely proportioned hull cut cleanly through the water under well-set sails. For all of the artistry that is brought to yacht design, however, there remains an inevitable connection between form and function. The myriad details which go together to achieve the finished form either work together or in conflict to create a design that may or may not be well suited for the sailing in any particular waters.

This balancing act between form and function is one which yacht designers wrestle with on a daily basis. Ideally, one could select a favorite designer and consider only boats carrying his (or her) name on the plans. Unfortunately, as in everything else, the real world doesn't work that simply. In the most fundamental sense, the yacht designer is the artist/engineer who creates a set of plans and then gives those plans to someone else to build the boat. Once those plans are out of his hands, however, the designer has little influence on how the lines he has put on that paper are transformed into a finished boat. The designer will usually specify a lay-up schedule and the execution of critical details. He will specify rigging, winches, through-hulls, hull-to-deck joint, etc. Whether those specifications are used in construction of the boat, however, is entirely the builder's decision. And yet, the designer's name goes on the label.

There are quite practical implications to this situation: In an interview some time ago, we asked Robert Perry what his name on a boat means. His answer: "It should mean that the boat will sail well." But that was all he would claim.

The credit line on a boat design goes to the person who draws the lines for the hull. As with specifications, the deck plan, interior accommodations, and rig all can be changed by the builder. The designer may even refuse to accept royalties for a particular boat. As long as the hull lines remain intact, however, the original yacht designer's name can be attached to the boat.

Sometimes the best source of information about a particular design may be but a telephone call away—to the yacht designer who drew the lines of the boat you are considering. Before depositing your coin in the telephone, however, there are some aspects of the yacht design business which, once considered, may be helpful in calibrating the designer's answers to your questions.

Boat designers generally work in one of three environments: They are in-house employees of major boat building companies; they are employees of a design firm; or they hang out their own shingles. Our experience suggests that being an employee of either the boat builder or a design firm may impose limits on what a designer can say about boats he has designed for his employer. However, we would expect most designers to say honestly whether they believe a particular boat of their design is suitable for the sailing you plan to do. They may also tell you whether the boat being produced uses their designs for the interior, rig, and deck plan—details worth knowing.

Boat designers also are sometimes victimized by builders who will purchase a design for a 34-foot boat and enlarge the plans to produce a 36-foot boat of the same lines. Or to reduce the drawings to make a 30-footer of the same basic lines. All three boats would carry the original designer's name, but the sailing performance of the long and short versions of the design may well suffer in the translation. In most instances, we would expect the designer to tell you if one of these situations exists with a particular boat carrying his name.

You can well wonder—as we have—why designers allow this kind of thing to go on. The fact is that designers carry very small sticks. They depend upon builders for their livelihood. Moreover, they are paid royalties on each boat the builder makes using their designs. For that reason, designers want their boats to sell. Some designers—we would think most—would tell you in answer to a direct question whether their specifications are being followed, if they know that fact. Unfortunately, designers have the same problem the rest of us have in telling how well a boat is constructed. As a result, it may take a year or more for a designer to learn that a builder is not following his specifications on some basic part of the boat, e.g., the hull-to-deck joint. In other instances, he may never know whether his specifications are followed. As a result, *even when a designer asserts he knows the quality and workmanship that go into one of his designs, you should not rely upon that assertion unless he personally oversees the construction of the boat.*

There is one other key element as well: Designers almost always have a client when they draw a new boat. Whether the client is an individual who wants a custom design, a boat marketer who wants a design which he'll farm out to a boat construction company to build while he markets the boat, or a builder who markets his own products—there is a client, and that client generally tells the designer what he wants. A person commissioning a custom design is having a personal boat drawn. The marketer has ideas about "what will sell" and those ideas provide the framework in which the designer operates. He may also have an image he wants to project with the boat. The boat builder has the same two considerations—a certain image associated with his boats, as well as ideas of what his customers want in a boat. The point is this: Seldom is the design of a production sailboat clearly intended for a specific use. Rather it represents a conglomeration of factors which include image, what clients believe will sell, the clients' own peculiar likes and dislikes, and, too often last, the designer's preferences.

All of these assertions, of course, have practical implications. Unless you opt for a custom boat, you probably shouldn't simply select a yacht designer. You need to select design features that are important to the sailing you want to do, and use them to help you sort through the potential candidates. Only when the list gets short, might you call the designers.

In the interim, the obvious challenge that we as boat buyers face is sorting out the variety of design details to ensure that the boats we finally focus in on are not only constructed but also designed for the sailing we want to do. It is a sorting out which extends far beyond the common conflict between the stereotyped heavy displacement cruising boats and light displacement flyers. But displacement is a good place to start.

DISPLACEMENT

We all know that if we fill a pot to the brim with water and then put a large rock in the pot, some of the water will overflow. It overflows because the rock takes up space in the pot formerly occupied by water. If we then collect the water that was pushed aside, or *displaced*, by the rock, and weigh it, we would determine the *displacement* of the rock. Boats work the same way. When a boat is put into the water, the hull pushes aside (displaces) a volume of water, the weight of which becomes a measure of the boat's displacement. The larger the *volume* of water that is pushed aside by the boat, the larger the boat's displacement. As a result, displacement can be looked at in two ways—as a measure of weight, and as a measure of the hull volume below the waterline. In summary, the heavier the displacement, the greater the volume of the hull below the waterline.

The discussion of displacement in sailboats pivots around one basic trade off: interior volume and comfort *versus* light air sailing ability and speed. The facts are these: If all other things are equal, a heavier displacement boat will have more room below the waterline than a lighter displacement boat. Moreover, the heavier displacement boat has a slower motion with the waves. Increased volume means larger storage capacity. It may also mean larger cabin areas and more headroom. Reduced wave motion translates into increased comfort, particularly on longer voyages. On the other hand, heavier displacement also means that more energy is required to get the boat moving because the hull itself is heavier and larger. If you will live aboard your boat at a dock, where you are concerned with space and comfort rather than with how hard or easy it is to get your boat moving, a heavier displacement boat may have much to offer. If, however, you plan on sailing principally in protected or semiprotected waters, where you do not need stores for more than a few days, and where how fast the boat sails determines your cruising range, a lighter displacement design may make more sense. In coastal and offshore waters, the choice is less clear. Some people argue that the time saved with a fast passage more than offsets the added fatigue from bouncing around in a light displacement boat. We find this argument particularly attractive for coastal cruising, where passages usually don't last much more than a week. We find it less compelling, however, for longer offshore passages. The cumulative fatigue over a 20-day passage from the movement of the light displacement boat can be quite large, particularly if several days of strong winds are encountered, outweighing the few days that may be saved by the faster passage. In foul weather, the heavier displacement boat may be easier to keep under control. The lighter displacement boat wants to take off at hull speed quickly, even under bare poles. Finally, for extended cruising, we prefer the greater stowage capacity associated with heavier displacement boats.

If the choice is between extremes of heavy and light displacement, the distinction between the two is rather clear even to the inexperienced eye. As a particular design moves away from the extreme, however, that distinction can blur. In this case a fairly simple formula which relates displacement to the load waterline (LWL) length can provide some guidance as to where a particular boat falls on the continuum from *light* through *moderate* to *heavy* displacement. The formula is easily worked with a calculator. As with any other such generalization, however, this formula should be used only as a guide. It will distort the results for boats designed with either unusually long or unusually short waterlines. The longer waterlines increase theoretical hull speed; shorter waterlines are designed to gain a rating advantage under measurement rules for racing boats. In either case, a difference of a foot or two will make a significant difference in the ratio calculated. Note also that displacement is expressed in metric tons, which is obtained by dividing the boat's displacement in pounds by 2,240.

$$\frac{\text{Displacement (lbs.} / 2{,}240)}{(0.01 \text{LWL})^3} = \text{D} / \text{L ratio}$$

To complicate matters, the notion of what constitutes a heavy or light displacement boat has been a moving target. In the 1970s and early '80s, a boat whose displacement/length ratio was between about 300 and 350 was said to be a "moderate" or "medium" displacement boat. One whose D/L ratio exceeded 400 was labeled "heavy." One whose D/L ratio was 250 or less was called "light" displacement. These categories were then refined further, with a D/L ratio between 350 and 400 characterized as "moderately heavy" and one between 250 and 300 as "moderately light."

More recently, those breakdowns have been adjusted to lend credibility to the trend toward lighter weight construction and improved light air performance. As a result, today you will find boats whose D/L ratio is 300 or above called "heavy displacement" boats; a D/L ratio between 200 and 300 is called "moderate"; from about 125 to 200 is "light"; and below 125 is called "ultra light" (Table 2). For most sailing, the new designations are fine. For coastal or offshore cruising, however, we prefer the older definition. We believe it provides a better description of cruising boat trade offs between performance and comfort.

There are other implications to differences in displacement in addition to the factors of comfort and performance. Heavier displacement boats generally require heavier standing rigging.

TABLE 2A. Characterization of Boats by Displacement/Length Ratio—1970s

Heavy	Moderately Heavy	Medium	Moderately Light	Light	Ultralight
>400	350–400	300–350	250–300	<250	—

TABLE 2B. Characterization of Boats by Displacement/Length Ratio—1980s

Heavy	Medium	Light	Ultralight
>300	200–300	125–200	<125

They may require heavier running rigging, larger sheet winches, heavier dock lines, heavier anchors, and heavier anchor lines or chain. Heavier displacement boats also usually require larger engines for satisfactory performance under power. They also require larger sails for satisfactory sailing performance—an important consideration because sails must be handled rain or shine and in heavy or light winds, usually by a smaller crew than is found on comparably sized racing boats. In this matter, too, a formula can be helpful:

$$\text{Sail area / displacement ratio} = \frac{\text{Sail area (sq. ft.)}}{(\text{Displacement*})^{2/3}}$$

This ratio is analogous to a horsepower rating in an automobile or boat. It provides a measure of the sail power available to move the boat. In general, for a boat to have reasonably good sailing performance, we'd like to see a sail area / displacement ratio in the range of about 16. If the ratio goes much higher than 17, the boat may need more crew than cruisers often carry. Much below 15 and the boat probably won't move in winds below 10 knots. In any case, a heavy displacement boat also needs a larger sail inventory, particularly for use in light air. Ketches and schooners also require a larger sail area than sloops, cutters, or yawls for comparable performance.

Key Exterior Design Features

While displacement describes bulk or weight, it usually isn't possible to separate consideration of displacement from the broad subject of hull shape. We like to look particularly at five key elements of hull shape: the length and depth of the keel; the rudder configuration; the bilges; the style of the stern; and the profile above the waterline. It is also important to consider cockpit design and the rig.

*Displacement is expressed in cubic feet, which can be obtained by dividing D in pounds by 64. (One cubic foot of seawater water weighs about 64 pounds.)

Keel Length and Depth

There is little question that the most effective keel from perspective of stability, upwind sailing performance, and maneuverability is a short, deep fin keel. The depth puts the ballast low. The vertical length of the fin provides the greatest lift. And the short fore 'n aft length of the fin reduces the surface area and provides a pivot for the boat to turn on for fast tacking. The helm, however, may need constant attention to maintain a course.

At the other extreme, a long, full keel whose profile flows out of the stem and reaches back to a "barn door" rudder beneath the stern provides almost opposite performance (Photo 25). The boat probably will not sail well upwind and it may well be necessary to backwind the jib to make it come about. But, it will be easy on the helmsman, with the boat tracking like the proverbial train.

Designers have made all degrees of compromise between these

25. The traditional flow of the keel from the stem leading to the barn door rudder yields a boat which should track like a train. At the same time, windward ability may be less than desirable and tacking probably will be slow—even difficult in some conditions.

BOAT DESIGN

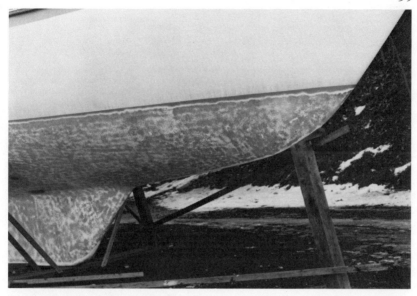

26. The cutaway forefoot of this moderate fin keel design provides good windward ability and easy tacking. The mottled appearance of the bottom results from sandblasting the gel coat to correct a beginning gel coat blister problem.

two extremes of keel shape. On modern boats with long keels, a frequent compromise is the "cutaway forefoot," in which the forward part of the keel is removed to improve both the upwind sailing and the tacking ability of the boat. In some designs, the cutaway is relatively little; in others the cutaway makes the forward part of the keel and hull bottom similar to that found in a fin keel design (Photo 26). The more closely the cutaway resembles the fin keel configuration, the greater the probable improvement in windward performance and ease of tacking.

Rudder Configuration

Just as there is agreement about the technically superior performance of the short, deep fin keel, there is also general agreement about the superior effectiveness and responsiveness of a spade rudder that is set well aft (Photo 27). That's why, in fact, so many high-performance sailboats for racing are designed with a high-aspect ratio fin keel and a spade rudder. The high-aspect

27. The spade rudder provides quick response and minimum wetted surface, but concentrates all of the rudder forces on the rudder post at the point it goes through the hull. There have been numerous reports of spade rudder loss or failure and increasing numbers of skippers of spade rudder designs are installing hardware for rigging emergency rudders before entering offshore races.

ratio means that the keel or rudder is deep compared to its length fore 'n aft.

Though used principally on boats intended for racing, spade rudders may be quite suitable for cruising boats that will be sailed in protected or semiprotected waters and whose owners want that kind of performance. We have serious reservations, however, about the suitability of spade rudders for coastal or offshore cruising. They are easily fouled by such obstructions as floating ropes in congested harbors, lobster pots, crab pots, and the live kelp found along many coasts. Spade rudder configurations also usually offer little or no protection to the boat's propeller. They are vulnerable to collision with floating debris. They may stall in severe conditions so that steering control of the boat is lost temporarily, quite likely causing a broach. And they may fail structurally (the rudder post may bend, for example) under the extreme forces generated in strong wind conditions.

Alternatives to the spade rudder configuration include the traditional rudder attached to the after end of a full-keel or the more modern skeg-mounted rudder (Photos 28 and 29). We prefer the latter. Setting the rudder well aft of the keel clearly

28. The traditional keel-hung rudder is typically a sturdy installation, but does not provide the steering power of a spade rudder despite its large size.

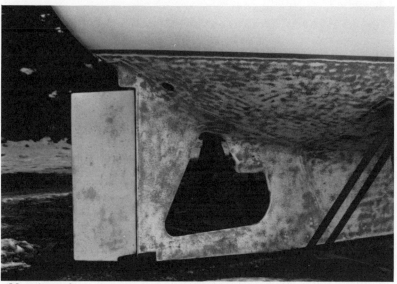

29. A modern skeg-mounted rudder can represent a good compromise between spade and keel-hung rudders. On this boat, a sturdy FRP strut between the keel and skeg further strengthens and protects the assembly while offering protection from fouling underwater lines—e.g., lobster pots.

30 and 31. The skeg of this heavily constructed cruising boat has been damaged (see close-up), apparently from groundings, but the rudder has escaped relatively unscathed.

provides a more powerful and more responsive rudder. At the same time, a full skeg can protect the rudder against impact with debris and strengthen the assembly significantly. The skeg also may greatly improve the boat's tracking ability and reduce or even eliminate the potential for the rudder stalling. The degree of benefit gained from the skeg, however, depends upon its design. A skeg which ends halfway down the length of the rudder, for example, can't provide the protection and strength of one which extends the full depth of the rudder (Photos 30 and 31).

The Shape of Bilges

One of the basic design elements common to light-displacement boats is a long, relatively flat, relatively shallow run of the bottom from stem to stern. This design feature has several purposes: It contributes to improved sailing speed in light air by

reducing the wetted surface area of the bottom, thereby reducing the friction between the hull and the surrounding water. It also enhances a boat's surfing ability—a great experience in a race when there is a large crew, but not so great for a couple alone on the ocean.

On the other hand, a long, shallow, flat run of the bottom severely limits stowage area within the boat. As a result, most light displacement boats have little stowage area beneath the cabin sole. Water tanks generally are placed under the settee benches because there is no room in the bilges, reducing available stowage area still more. The fuel tank often is put somewhere under the cockpit area. There may well be no deep sump for bilge water, making a small amount of water in the bilges often a major nuisance—possibly even a dangerous nuisance if it slops onto the cabin sole as the boat heels to a strong puff of wind, making footing flippery.

In contrast, a moderate-to-heavy displacement boat has fuller, deeper bilges. Probably the water and fuel tanks are below the cabin sole. There is usually a sump for bilge water which keeps the water below the cabin sole except, possibly, in extreme weather. There may also be stowage area beneath the sole for canned goods, spare anchors, and chain. The trade off is a reduction in light air performance, and a reduced likelihood of surfing on a run in strong winds.

If one is sailing on the ocean, either coastwise or offshore, the advantages of tankage below the cabin sole, of increased stowage space, and of a deep sump for bilge water may well be important. They may also be important in any waters for a live-aboard boat; few who live aboard their boats ever have enough stowage space. For daysailing or week-ending, or for cruising for just one or two weeks at a time, however, the advantages of fuller bilges and heavier displacement may become less important and the performance penalties progressively more important. It depends upon the needs, likes and dislikes, and wants of individual boat owners.

Stern Configuration

There is continuing discussion over what shape stern is best for ocean sailing. Although our own bias is for a canoe shaped stern, we're not certain whether that bias results from our sense of beauty and symmetry, or from a perception that the canoe stern really is better in foul weather because it tends to part following seas. Aside from that debate, however, there are some readily apparent practical considerations about the shape of the stern when selecting a boat. For example, a canoe stern generally provides less cockpit and stowage space. For an offshore boat, that's not all bad. For a boat likely to carry guests, that impact on cockpit size and stowage is a potential problem.

Boats with reverse transoms present a different circumstance (Fig. 12). The reverse transom represents an effort by design-

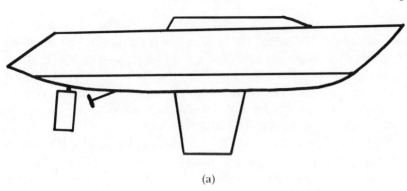

(a)

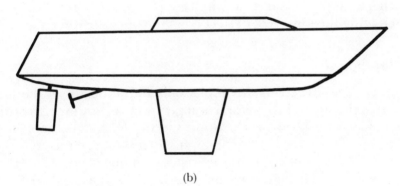

(b)

FIG. 12. The static waterline length and hull configuration of boats (a) and (b) are identical, except for the transom design. When boat (a) moves through the water, the hull wave formed will build up under the extended stern, lengthening the effective waterline length. Heeling will cause the same effect. All other factors being the same, boat (a) will be the faster of the two.

ers to maximize the effective length of the waterline, thereby improving the boat's sailing performance. One way to extend the waterline is to make the boat longer in the stern section. If that's all that was done, however, the change would penalize the boat's weight aft. An alternative is simply to extend the length down near the waterline, but omit most of the remaining stern section by cutting a diagonal from the tip of the stern forward to a point on the after deck. We believe the reverse transom style presents a disadvantage for extended cruising: The stern ends in a rather sharp edge where the bottom of the transom joins the hull just above the waterline. This increases risk of damage from bangs by dinghies, other boats, and pilings or bulkheads in marinas or at public docks.

A traditional stern with a long overhang increases the potential for hobbyhorsing in certain wave patterns. The problem arises because the overhanging stern usually provides lazarette space for a substantial amount of gear. The weight of that gear is well aft of the boat's center of gravity, creating a long lever arm which tends to make the boat rock fore 'n aft, and the hull shape does little to dampen the motion. The problem may be made worse if the bow has a similar overhang so that anchors, chain, and—possibly—a windlass put weight well forward, creating the same lever effect from the opposite direction.

Freeboard

Technically, freeboard refers only to the height of the hull up to the sheer. From a practical perspective, however, the concept of freeboard needs to include the height of the overall profile of the boat. In general, high freeboard tends to result from the designer's effort to provide standing headroom. However, freeboard also means windage, and as freeboard is increased, the effect of the wind blowing on the sides of the hull and cabin house is also increased. The practical result is that the wind blowing against the side of the boat pushes it to leeward. The more freeboard for the wind to blow against, the greater the leeway. This is a particularly important consideration in trying to beat upwind efficiently. It is also an important consideration when maneuvering in close quarters under power. In any sailboat—particularly at slow speeds—the force of the wind on the bow may quickly become more powerful than the effect of the propeller and rudder, blowing the boat out of control until the skipper can either turn the bow away from the wind or give more power to increase boat speed enough for the rudder to overpower the wind. The greater the freeboard, the more difficult such maneuvering may become. As a result, given a choice for almost any use between two boats—one with a high freeboard, the other with low freeboard—that are otherwise equally well suited to our needs, we usually would choose the boat with the lower freeboard.

Cockpits

In addition to details of locker design, the companionway hatch, and the bridge deck discussed earlier, two key factors of cockpit design are size and elevation above the waterline. The principal importance of size is that a cockpit should be large enough on boats intended for dockside living or sailing in protected or semiprotected waters to accommodate the number of people likely to be on the boat. Alternatively, the cockpit of a boat intended for coastal or offshore sailing should be small enough that a wave flooding it would not imperil the boat's stability. Such a cockpit probably is too small sometimes to seat guests comfortably, but the safety factor is more important than the occasional inconvenience of its small size. In any case, a boat for offshore use should have large, unobstructed cockpit drains with the hoses to the sea cocks crossed to prevent water backing up through the drain hoses into the cockpit when the boat heels. Ideally there would be four 1½-inch drains; if that is not possible, there should be two 2-inch drains.

The question of elevation is one of comfort more than of safety. In some center cockpit boats, the cockpit is quite high. That's not a problem for dockside living, or for use in protected waters. It may be a problem in semiprotected or ocean waters because the height of the cockpit magnifies the motion caused by waves. In rough weather, likelihood of seasickness may be increased, even for people who normally are not prone to motion sickness.

Rig

High aspect ratio (tall mast, short boom) sloops offer the most efficient sailing rigs and are probably the rig of choice for most boats of about 35 feet or less that will be sailed principally in protected or semiprotected waters. Depending upon the vigor and size of the crew, the sloop rig may be workable in boats up to about 40 feet in length. At 40 feet, however, the sails are pretty big. The sailcloth on the mainsail is probably heavier on the larger boat as well. For these reasons, one is well advised to look at split rigs above 40 feet. Usually these are ketches, but a few boats may be offered as schooners or yawls. In our view, a 40-foot boat is also getting large for a cutter if the boat is for a couple. For coastal or offshore cruising, cutters and split rigs have one great advantage: multiple sail configurations. When the wind pipes up, the cutter can reef the main and drop the jib, sailing on staysail and reefed main. With a double reef in the main and the staysail flying, a well-found cutter can keep sailing efficiently in quite strong winds. Ketches often can sail efficiently under jib and jigger in stronger winds. The advantages offered by cutters and split rigs, however, do not mean that sloops are unsuited for coastal or offshore sailing. To the contrary, the sloop rig is just as efficient offshore as it is inshore. It just doesn't offer quite the versatility in dealing with storm winds.

With any rig, we suggest looking carefully at how and where the mast is stepped. A deck-stepped mast, for example, should have a sturdy compression post between the underside of the deck beneath the mast and the keel. We prefer stainless steel compression posts. We have seen wooden posts and bulkheads used for this purpose be compressed in use until the cabin top deflects. Stainless steel also is preferable to aluminum because the bottom of the post may stand in bilge water at times.

If a mast is stepped through to the keel—or a compression post to the keel—we suggest being certain that it is stepped on the ballast keel. In some of the modern fin keel designs, the leading edge of the fin is so far aft that the mast is stepped onto the hull, not the keel. We know of one boat from a well-respected builder that began leaking as it pounded down the lower Chesapeake Bay at the start of the Bermuda race recently because the combined force of the pounding on the hull around the mast step and tension on the backstay cracked the hull in the area of the mast step. In that case, leaking was stopped by easing pressure on the backstay, but the lesson should be obvious.

Certainly for coastal or offshore sailing, and preferably for sailing in semiprotected waters, the mast or mast compression post should be stepped on the ballast keel.

KEY INTERIOR DESIGN FEATURES

In shifting attention from the exterior to the interior, it is often tempting to relax and to think about comfort and accommodations. It is a temptation to be resisted. Safety is at least as important as comfort and accommodations and may often be the first consideration.

Cabin Space

The differences in safety considerations in using a boat for dockside living, for sailing in protected or semiprotected waters, or for coastal or offshore cruising often are differences in detail that reflect the specific conditions the boat will encounter. When thinking about the space below decks in a boat intended for dockside living, a designer does not have to be concerned about the cabin being too large, or about how far crew members are at any time from a handhold. Once the boat moves away from the dock, however, distances between handholds become increasingly important, as does the amount of open space in the cabin. A boat used only in protected waters, for example, can still be rocked violently by powerboat wakes. When that happens, the people down below need a handhold immediately available. Even moving around in a steady boat that is heeled over 20 degrees as you sail along requires something to hang onto just to move around. Most of us, at least, are not accustomed to walking with a 20 degree list.

Cabin size also is a safety consideration once a boat leaves dock. A sailboat under power or sailing downwind in light air is susceptible to potentially violent rocking from the wakes of 25- to 40-foot powerboats speeding past. A person caught unawares down below and not holding onto something literally can be thrown from one side of the cabin to the other. The larger the open width of the cabin, the farther there is to fall and the greater the potential for injury. The importance of both handholds and cabin size to crew safety increases as the boat moves into semiprotected and then into ocean waters, because boat motion becomes even more complex and less predictable.

Decor

In the 1970s, fiber glass interiors developed a bad name—the "plastic" or "Clorox-bottle" look. One result was an effort to hide the fiber glass behind fabrics, veneers, and wood. Moreover, as the sailboat business grew larger and more competitive, interior designers were brought into the act to strengthen the competitive marketing advantage of various builders. The result is that by the advent of the '80s, the interior decor of most sailboats had lost a certain association with function they once enjoyed.

For daysailing, or even cruises of a few days or weeks, the interior decor is relatively unimportant; the crew is likely to be more caught up in the cruise than in the decor. For living aboard, however, or for offshore passage making, the decor can be quite important, particularly as it affects the crew's mood when, for example, they are effectively trapped below by two or three days of grey, rainy and, possibly, rough weather. For this reason, we prefer boats whose interiors are light and airy.

Location of Sinks

Sinks are supposed to drain water out of the boat. If they are positioned badly, they can do just the opposite, causing serious flooding instead. Such problems are unlikely in a boat used for dockside living. As boats move from protected to semiprotected waters, and then to coastal or offshore waters, however, the importance of the sink location increases because the boat is farther away from help should a problem develop. The key, as we've mentioned more than once in earlier chapters, is that sinks should be positioned along the centerline of the boat. They should also be well above the waterline. In this way, no matter how far the boat heels over, the sink will remain higher than the outside water level and sea water cannot siphon into the boat.

But that's how designs *should be* drawn. Because interiors usually are designed for marketing purposes rather than to accommodate the idiosyncracies of sailboats, sinks—particularly in the head—are often located well outboard. In such designs, sea cocks for shutting off the sink drains should be immediately and easily accessible so that they are kept closed whenever the sink is not being used.

Head (MSD) Location

There are as many opinions as there are sailors as to where the head should be located. Our only concern is that the location make sense in each particular boat. However, apart from location, there are several important considerations in head design, no matter where on the boat the head is located. The first four apply to all boats; the last two become increasingly important as the use intended for a boat reaches out from protected waters toward the ocean. (1) The sea cocks for the head intake and the outlet through-hull fittings should be readily accessible. (2) The head seat should be located above the waterline to reduce the potential for water siphoning back into the boat if sea cocks are left open inadvertantly. (3) The head discharge hose should be looped up to deck level and a vacuum relief valve placed at the top of the loop to help protect against water siphoning back into the boat. (4) The sea cocks for intake water and outlet should be well separated, with the intake located either two or three feet forward of or on the opposite side of the keel from the outlet through-hull. (5) The head seat should be where you can wedge yourself against pitching and rolling. (6) Firm handholds should be placed on both sides of the seat.

Galley

On any boat, the potential for fire caused by careless use of the galley stove must be minimized. This means keeping curtains away from the stove's flame. It also means lining the stove compartment and, possibly, the inside of the cabin house or deck above the stove with stainless steel. Once past concern about fire prevention, however, the design of the galley on any boat is a compromise between convenience and safety. At the dock, convenience can take precedence, much as it does in our homes. Once a boat moves away from the dock, however, convenience should begin to lose out to safety. For example, the galley should be located off to one side or the other so that it is not part of a high traffic area. If the galley will be used at all while under sail or power, it should be designed so that the cook can wedge himself in when the going gets rough, leaving both hands free to work. The most common arrangement is to form a U-shaped galley using the sink, stove, and icebox as the three sides of the U. However, there must also be a sturdy bar to keep the cook from being thrown into the lighted burners. There should also be provision for a strap the cook can lean back against to keep from being thrown backwards out of the U.

The icebox should be top opening to help keep the chill inside and to keep the box contents from piling out into the cabin when the box is opened. Insulation around the box should be at least four inches thick all around. Ideally, an alumized Mylar film is placed as a vapor barrier between the icebox liner and the foam insulation. Large is not necessarily good in iceboxes. Six cubic feet is a big icebox, providing plenty of room for ice and food. Refrigeration is your option.

In addition to the location of the galley sink, the design of the sink is important. If cooking while sailing, the sink often is the only place a hot pot can be put safely once it is removed from the gimbaled stove. For this reason, we prefer sinks that are at least 10 inches deep and large enough to hold our largest pot—a pressure cooker. Double sinks also are useful, though not a necessity. If the choice is between one good-sized, deep sink and two smallish sinks, we'll take the single, good-sized sink.

Tables

Once it is established that the table in the main cabin is large enough and well enough placed so that it will be comfortable in use, the principal requirement of any table on a boat that will be taken away from the dock is that it be sturdy. We are not fans of tables that fold up against a bulkhead. We believe they are accidents waiting to happen when they are folded up, and they are only poorly anchored when in use. Ideally, the table would be tied to a bulkhead at one end and to a post running from cabin sole to overhead at the other. In that way, the table has a chance of absorbing the impact of someone falling against it in a seaway without collapsing. The sole to overhead post also provides a convenient handhold right in the center of the cabin.

Berths

One of the problems boat buyers face today is that most boats are designed either with only one or two berths, or with enough berths to open a boarding house. Moreover, some boats come with what appear to be "designer" berths. Again, if a boat will be used at the dock, the question of berths is relatively simple: There need only be sufficient berths with sufficient privacy to accommodate a certain number of people comfortably; moreover, they can be in any shape or configuration desired. There need be no connection between form and function. If the boat will be used for short cruises in protected and semiprotected waters where it will be at anchor every night, the number of berths is again the principal consideration. In this instance, however, you may want to consider whether settees or dinette seats should be used as berths. The argument for using them as berths is that you can accommodate more people. The argument against using them is that once they are made into berths, that's all the space can be used for. If you need to sleep four people, however, and can accommodate them by putting two in a vee-berth forward, one in a pilot berth, and one in a quarter berth, anyone can go to bed at his or her convenience while others in the group continue to use the settees or dinette seats.

Once you are planning coastal cruising or offshore passages, the business of berths becomes serious. First, each berth must have something to keep the occupant from falling out of bed. It may be a lee board—a piece of wood running the length of the berth and extending five or six inches above the mattress, in effect serving as a guardrail. As an alternative, it may be a lee cloth—a length of canvas fastened to the frame of the berth and tied to the overhead to serve as a guardrail. Second, your berths should be narrow, preferably in the range of about 22 to 24 inches. That is wide enough for comfort, but narrow enough to avoid being rolled around. For practical purposes, this means that double berths are often useless offshore unless there is provision for a lee cloth down the middle. Usually, one wants to wedge his or her body against the hull, a partition, lee board or lee cloth to keep from being tossed or rolled around by the action of the waves. Two people attempting to sleep in the same berth usually will be rolled into each other, possibly painfully.

There is often argument made about the utility of a vee berth. A proper vee berth—one that is long enough and wide enough for two people to sleep comfortably—makes a nice double berth for use inshore. Offshore, it presents the most difficult sleeping situation because forward is where most of the motion is. However, if there are alternative single berth accommodations for use offshore, the forward vee berth can be used effectively as a sail locker while the couple uses the single sea berths aft for the duration of the passage.

Water Tanks

If a boat will be used at dockside or in essentially local waters where the water tanks will always be filled from the same, reliable water source, one can get by with a single water tank. For cruising longer distances, where water will need to be replenished, or where a new supply of fresh water may not be available for two or three weeks, we believe it imperative to have two or more water tanks. Moreover, the fill and draw systems should be arranged so that each tank can be filled and the water drawn from it independently. This accomplishes two purposes: If one tank becomes contaminated, it can be isolated from the others. In addition, a leak in the water system will only drain the one tank being used. If you have only one tank and it leaks, you will have lost all of your water; if you have two or more tanks and they are isolated from each other, only the water in the leaking tank will be lost. Moreover, as you use water, emptying one tank reminds you that you have used that part of your water supply. With only a single water tank, it is too easy to run out of water at an inconvenient time.

Stowage

The stowage needs for any particular boat differ significantly depending upon the use intended. A boat used for dockside living, for example, has stowage needs more nearly like those of an apartment. A boat for daysailing or short cruises needs only to accommodate clothing, bedding, and food for a few hours or days. A boat headed offshore may need to be independent of land for 30 days—45 when you add a safety factor. Stowage spaces on most boats produced today seem to reflect the needs of the dockside sailor more than those of the sailor who wants to venture away from the dock. For example, designers tend to fill boats with hanging lockers (closets) and drawers. Where stowage is not a concern, and where you need to wear clothing that should be kept on hangers to look nice, hanging lockers have a valid function. On many—if not most—sailboats, however, hanging lockers have little real purpose and represent a relatively inefficient use of stowage space. The same clothing that you would normally put in a hanging locker generally can be stowed more efficiently by being rolled and kept in a bin or on a shelf. The exception is the seldom-found hanging locker for foul weather gear, hopefully located right by the companionway so that foul weather gear can be stowed before spreading its water throughout the cabin.

Drawers are another stowage system often overdone on sailboats. Drawers are convenient, but they are also space wasters. Where stowage space is important, it may be worthwhile considering replacing drawers with lockers which provide added shelf space. The shelves in these (and all other lockers) should have fiddles at least an inch high to keep things from sliding out when the locker door is opened.

We sometimes talk about design features as if they were abstractions. Indeed, when design is dissected into individual components as we have done in an effort to relate single features to the use intended for a boat, perhaps we are dealing with abstractions. But every once in a while, we run across a comment in a book or magazine article which brings these abstract details of design (or of construction) back to the reality of the sea more effectively than we are able to do on the basis of our own experience. We found one such comment in *Come Aboard* by Eric Hiscock—a comment which, in a homely manner, captured our feelings about the importance of small cockpits and bridge decks on boats for offshore sailing: ". . . Suddenly, the sea around us became steep and confused, and one heavy crest built up so high it lost its balance and fell into the cockpit, from which it proceeded through the open companionway into the sleeping cabin and chart space."[1] The incident is particularly impressive when you realize that Hiscock's boat at that time was the 49-foot center cockpit ketch, Wanderer IV.

---

1. Eric C. Hiscock, *Come Aboard* (New York: Oxford University Press, 1978), p. 20.

III

Making a Decision

Perhaps the greatest difficulty in making a selection from the finalist boats is keeping options open by recognizing the emotional preference likely to develop. Often there are sound—even if subconscious—judgments supporting that emotional response. Sometimes, however, emotional reactions to a boat can lead toward bad decisions. It is this potential the boat buyer must be wary of. Having gone this far rationally, you can have confidence in your ability to make a good selection as long as you evaluate each boat honestly and with conscious awareness of any emotional biases that may have developed.

8

The Second Time Around
A Closer Look at the Leading Ladies

After months of studying and looking, you will probably have narrowed your list of boats down to a few leading candidates. This is a good time to go back to boat dealers because now you will be looking at their boats much as single men and women often look at each other when they go out together for the second time. The first time around, you were scouting the field. This time you are considering whether you want to get serious.

So you have two basic objectives. The first is measuring the boat much more carefully against a set of priorities you have now established. The second is taking the measure of the dealer. He's like in-laws: they come with the marriage; he comes with a boat.

MEASURING THE BOAT

In establishing the priorities for a boat, we all end up making a series of compromises between what we know intellectually is "best" for the sailing we want to do, and what we "want." Hopefully, however, compromises will be in details, not basics. There are fundamental considerations relating to use which are best held above compromise or you are likely to be unhappy with your selection (Table 3). For that reason, it is important to have those fundamental priorities firmly established in your mind before beginning the final selection process.

Table 3. Fundamental Priorities for Five Boat-Use Categories

Dockside	Protected Waters	Semiprotected Waters	Coastal Waters	Offshore Waters
Safety —Fire —Sea cocks —Lifelines & ladder	Safety —Rigging —Lifelines —Handrails —Rounded corners —Sea cocks	Safety —Rigging —Reefing —Sea cocks —Lifelines (sturdy) —Handrails —Smaller cabin areas —Safe galley —Rounded corners —Engine access for trouble shooting	Safety —Hull strength —Rigging —Reefing —Sea cocks —Lifelines (sturdy) —Handrails —Bridgedeck —Small windows —Mast on keel —Handholds below deck —Smaller cabin areas —Safe galley —Rounded corners	Safety —Hull strength —Hull/deck joint —Rigging —Mast on keel —Sea cocks —Lifelines (sturdy) —Small windows —Bridgedeck —Small cockpit —Handrails —Handholds below deck —Smaller cabin areas —Safe galley —Rounded corners —Access to all systems for repair —Sea berths
	Sailing Performance —Speed —Light air —Close winded	Sailing Performance —Speed —Heavy-weather ability	Sailing Performance —Windward ability —Heavy-weather ability —Easily balanced	Sailing Performance —Windward ability —Heavy-weather ability —Easily balanced

Dockside	Protected Waters	Semiprotected Waters	Coastal Waters	Offshore Waters
Comfort –Space –Accommo- dations –Insulation –Ventilation –Lighting		*Comfort* –Fiddles –Sturdy latches	*Comfort* –Ventilation –Sea berths –Fiddles –Sturdy latches	*Comfort* –Ventilation –Fiddles –Sturdy latches
Storage –Everyday items –Seldom used items				*Storage* –Multiple water tanks –Room for 40-day supply of food –Spares and emergency repair items –Emergency stores

Dockside Living

We would set three coequal priorities for dockside living. Listed alphabetically, they are comfort, safety, and storage space. Sailing performance and motoring performance, on the other hand, would be of relatively little interest since the boat will rarely leave the dock—and then probably in hand-picked weather.

Comfort involves a number of parameters. Some are fairly obvious: The arrangement of the accommodations to ensure desired privacy; the arrangement, design, and construction of seating so that you and guests can sit comfortably and talk for several hours; lighting that will let you read comfortably after dark or on gray, rainy days almost anywhere you might want to sit or curl up with a book; and the design of the galley to provide work space for the occasional fancy meal you'll want to prepare.

Other parameters of comfort may be more subtle. While it

is obvious that a boat for dockside living requires a substantial number of opening hatches and portlights for comfort in warm weather, for example, it may be less obvious that boats intended for dockside living need ample dorade boxes with large cowl ventilators to provide fresh air below in foul weather. Obstructions to ventilation through dorades or portlights also may be less than obvious; opening ports shielded from gentle winds by high bulwarks, for example, do little for ventilation at a dock.

If the boat will be used in cold weather, insulation may be important to eliminate problems of condensation, particularly dripping from the overhead. Insulation may also be a consideration if the boat will be air conditioned. A cored hull and deck may provide the required insulation, but not necessarily. Areas constructed without the core material for attaching deck or hull fittings may become centers of condensation during cold weather.

The safety considerations are greater in number. Four principal safety concerns at dockside are fire; sinking; drowning by falling overboard; and electrolysis, with subsequent failure of underwater fittings, including through-hulls. The most likely source of fire is the galley stove. For that reason, the area around and above the stove should be shielded from flame. Curtains are pretty, but they may also be flammable and should not be where flames can reach them. Under no circumstances should curtains near the stove be loose at the bottom where they can hang out or be blown out over the stove. Additionally, a fire extinguisher should be immediately available. If propane is the fuel used in the stove, the fuel tank should be mounted either on deck where leaking gas cannot run down into the boat, or in a separate locker which drains overboard.

Sinkings at dock usually are the result of a failed hose or hose connection. This means all through-hull fittings near or below the waterline should have shut-off valves, preferably sea cocks. All hoses connected to through-hull fittings should have two hose clamps on each end connection.

The key to preventing drowning is to keep people onboard with sturdy lifelines and easy access from the dock to the deck. However, even boats for dockside living should have provision for a quickly accessible boarding ladder. Particularly if the water is cold, or the victim has been drinking, it may be very difficult to get someone out of the water without the boarding ladder.

Stray electrical current from nearby boats or poorly grounded shore power circuits should be guarded against by providing an effective, common ground for all through-hull fittings and the electrical system. This will help prevent electrolysis and possible failure of through-hull fittings.

Storage space is perhaps more important than many people realize—until after they've bought a boat and run out of space. The key is having not only enough storage space for routine items such as clothing, bedding, towels, and food, but also for the myriad items you will gradually accumulate in your daily living. For some personal belongings, storage space may mean shelves where items can be put on display. For others, it may mean space under the cabin sole or under berths where belongings can be put away for indefinite storage.

Protected Waters

Our priorities for a boat used only in protected waters are safety and sailing characteristics. Storage capacity—important for dockside living—is less important for this boat unless you are planning cruises longer than one or two weeks. The interior arrangement and engine accessibility also would be relatively low on our list of priorities because, within reason, we could adapt to whatever features the boat provided for the short cruises we'd be making.

In the area of safety, we would look first at the rigging. We are less concerned about hull and deck construction, windows, and cockpit size because this boat won't be running into heavy seas. It may run into thunderstorms, however, and so the vulnerable area is the rigging. This means paying attention to chain plate installation, the stemhead fitting and the fittings used to anchor the backstay. It also means comparing the size of stays, shrouds, and winches when looking from one boat to another, and checking for backing plates behind all deck fittings, including winches.

Lifelines are desirable to help keep crew and sails on deck. Handholds or handrails are needed both above deck and below to keep the crew from falling. In the cockpit as well as down below, we'd look out for sharp edges and corners—hoping we wouldn't find any. We'd want shut-offs (preferably bronze sea cocks, though we'd accept plastic sea cocks in preference to gate valves) on all through-hulls that might be underwater when the boat is under sail. We'd be happier, however, with shut-offs on all through-hulls regardless of their location.

Sailing characteristics are particularly important for this boat. How well the boat sails, for example, will determine how far you can range in a day, or on a week-end. A boat which sails poorly will quickly be a source of frustration. Also, if weather does begin to deteriorate, a boat that sails well is more likely to get you into an anchorage or marina before the worst weather strikes.

Semiprotected Waters

As the potential for encountering rougher water increases, so too does the list of priorities. Moreover, we are now talking about longer passages—thirty to fifty miles—and six- to ten-hour sails. The weather may be very different at midday than in the morning, and different still by evening. The short chop of many of our larger bodies of semiprotected water may mean a lot of pounding, with the result that hull integrity is now a serious consideration. We would require readily accessible bronze sea cocks on all through-hulls and two clamps on all hose connections. Rigging must be sturdy; not only does it need to hold up under sudden strong winds, but also under sustained strong winds and the jarring forces of a beat to windward against a choppy sea. Easy reefing now is important. Preferably, the boat will have double lifelines. Whether single or double, they must be sturdy. Backing plates on all deck fittings are critical.

On the interior, we'd begin to worry about ricocheting around the open cabin; for that reason, we'd be looking for smaller open areas. Sturdy handholds are a necessity, as are rounded corners. Storage lockers and drawers must have strong, positive latches so they won't pop open as the boat pounds through choppy seas. Tall fiddles are a must on all shelves, counters, and table edges. The galley stove should be gimbaled if you have any aspirations of cooking while under sail. A sturdy bar should be put across the front to keep the cook from falling against the stove. We would also begin to consider seriously now the matter of engine accessibility. Help is potentially far enough away that it is important to be able to add oil, change filters, bleed the fuel system and replace a water pump impeller or an alternator belt without being a contortionist.

Coastal Waters

Our oceans and the Gulf of Mexico are serious waters, even when sailing coastwise. Not only can winds whip up very large and heavy seas, but floating debris ranging from huge logs or railroad ties to shipping containers are very real hazards. Closer to shore, different hazards are presented by sand shoals, coral, and rocks. Moreover, wind direction and the land restrict where you can go. As a result, the safest course may be to head offshore if foul weather arises, an eventuality which means your boat should be capable of handling those conditions and have space for enough water and stores that you can spend an extra day or two away from land without hardship. What all of this means, of course, is that a boat planned for coastal sailing should be almost as serious as one planned for crossing oceans. It should also sail to windward reasonably well, even in heavy weather.

If hull integrity is a serious consideration in semiprotected waters, it is critically important in coastal waters. This means a strong laminate and hull-to-deck joint, bulkheads well bonded to the hull or firmly attached to the hull liner, and readily accessible sea cocks with double hose clamps on all through-hull fittings. The integrity of the deck and cabin house are equally important. Windows should be relatively small or protected by strong storm shutters. Hatches must be strong and able to be securely fastened from the inside as well as the outside. Rigging must be stronger still. Ventilation through dorade boxes is very desirable if not a necessity.

Preferably, the mast will be stepped on the keel. If a ketch, at least the mizzen should be keel-stepped. In any case, rigging should be notably sturdy. Easy reefing and good sails also are a necessity; the engines in sailboats of the size ranges we're considering are virtually useless in heavy weather. Double lifelines and strong lifeline stanchions are a requirement, with no compromise permitted. There should be readily accessible and strong handholds on the cabin house and / or deck to facilitate moving around in rough seas.

Down below, we would look very carefully at the availability of strong handholds and the size of open spaces. Visual space is welcome; physically large spaces are unwelcome because of the difficulty they present in moving around in foul weather and the risk of falling the longer distances. In addition to such details as high fiddles and secure latches for cabinets and drawers, sea berths are now important. One must have a sufficient number of sea berths to accommodate at least the off-watch crew.

Offshore Waters

The basic differences between boats that are suited for coastal cruising and those that are well suited for crossing oceans are differences of degree, particularly in the amount of stowage space and water tankage, the strength of the hull laminate, strength of the rig, and the accessibility of all systems. The increased need for stowage space and water tankage for offshore cruising is easily understood. You simply need to carry more water and stores for a 20-day passage away from land than for a two or three day trek down the coast. The increased importance of hull strength and accessibility are perhaps more subtle, but both are related to safety. The farther you go from shore, the greater the need to rely upon your own resources when problems arise. For example, there are a number of boats constructed using hull liners and relatively thin hull laminates that we would sail coastwise because the rest of the boat is well enough executed to make it safe for coastal cruising—unless it had the bad luck to run hard into a partially submerged shipping container and develop a serious leak. In that circumstance, if we were unable to stem the leak, rescue would be reasonably close at hand. We would probably not choose those same boats for crossing an ocean, however, because the liners make repair much more difficult and the opportunity for rescue is considerably less one or two thousand miles offshore than it is just ten or twenty miles from the coast. The same difficulty of rescue offshore also mandates easy access to all of the boat's systems as well as to its through-hull fittings so that any failures can be located and repaired quickly to ensure the safety of the boat. Again, a boat constructed using a hull liner which allowed good access to through-hull shut-offs, but not to electrical and plumbing systems, might be suitable for coastal cruising but would probably not be the boat of our choice for offshore use. In addition to sturdy chain plates, headstay and backstay fittings, and rigging, we would prefer a keel-stepped mast on a boat for offshore waters. With all other things being equal, a keel-stepped mast is simply stronger than one stepped on the deck.

Going through the Boat

This is the second time around. Unlike the earlier look, when you were screening a large number to narrow the field to a select few for closer examination, you have the time necessary to look into every nook and cranny that you can get to. You will be using the same kind of methodical approach and looking at the same kinds of details provided in Chapter 5. By now, however, hopefully you have developed a detailed list of priorities which includes both design and construction details. Some of these elements may be identical to those suggested for the 15-minute screening process. Others, however, will probably have emerged from looking at the variety of boats and from talking to other sailors throughout this selection process. Some ideas may have emerged from the pages of this book. In any case, this list represents your yardstick—the device you will use to measure each of these boats. For it to be useful, it must be as complete as you can reasonably make it; the alternative is a less-precise measurement. Moreover, this priority list must be applied evenhandedly for each of the finalists.

Whatever your list of priorities, the key is in being methodical as you go over each of these finalists thoroughly. We suggest first that you carry a checklist of your priorities and that you work from the same checklist in looking at every boat. We suggest also that you begin on the outside, stepping back to look at the boat from the ground or dock and moving slowly from bow to stern with your examination and *making extensive notes about different features, noting both positive and negative features.* From this vantage point, for example, you can make notes about any of the through-hull fittings above the waterline or about running light configurations. If the boat is out of the water, you can also get a good look at the keel and rudder configuration, the protection offered the propeller and shaft, and through-hull fittings located below the waterline. As you move from the ground or dock to the deck, and then down below, the same procedure should be followed—moving methodically from bow to stern, measuring the boat against the list of priorities that you have developed, and documenting that measurement in detail with your notes.

MEASURING THE DEALER

The dealer is almost as important as the boat. In fact, the difference between a good and a not-so-good dealer can make or break the experience of buying your boat. It can also cost you a lot of money. The dealer is your link with the builder. He is the person who will take your money. He is the person responsible for commissioning your boat, and for seeing that any warranty claims are taken care of after the boat has been delivered.

Your job is to "interview" the dealer to learn enough about him to determine whether he (or she) is a person you want to do business with. We have a preferred approach to these "interviews," and that is to avoid letting anyone know how much studying we've done. In fact, we have been known to appear quite ignorant about boat design elements and construction. Later as we unveil whatever knowledge or insights we do have, we listen carefully to see if his story line—and specific answers to questions—remain consistent throughout our "interview."

Willingness to Help

One question you can answer quickly is how much help the dealer will be in providing information you need to make your decision. For example, a dealer may not have a hull lay-up schedule or core samples for the boat you are interested in. He also may not know what types of resins are used in the hull and deck. But he can get that information and, perhaps, a core sample. Moreover, a dealer who is willing to put out that effort on your behalf is one who creates good feelings. A dealer who tells you—as more than one has told us—that lay-up schedules are worthless because you can't tell anything from them is blowing smoke. Among other things, a lay-up schedule purported to be for the boat you are looking at gives something concrete that you can compare with other boats you are looking at. If you can also obtain a core sample, and find out what part of the hull it came from, you can have the sample burned. After burning, only the layers of fiber glass reinforcement will remain, making it possible to check the lay-up in the sample against the lay-up schedule. Such testing may sound extreme, but laminations do get left out. For example, we know personally of an instance involving a reputable builder in which a missing laminate in the hull made a boat list significantly. The omission was identified by burning cores removed for through-hull fittings to count the laminations. Once the problem was identified, the buyer was given a new boat; the boat lacking the lamination was later sold as a "second."

Demonstration Sail

If sailing characteristics are important for the use you intend for your boat, we would walk away from a dealer who won't provide a demonstration sail without requiring a deposit and purchase contract. However, you should not be surprised if a dealer wants to charge a reasonable fee for the demonstration sail. Usually that fee will be applied against the purchase price if you order a boat. In any case, the fee is asked to cover the cost of insurance and the time of the people who take you sailing.

Customer List

One of the best ways to qualify a dealer is to talk to his customers—sailors who have purchased a boat from him that is similar to the boat you are considering. If the dealer won't give you more than one or two names, we'd want to know why. We wouldn't accept the argument that he doesn't want to impose upon his customers by giving out their names. In our view, that's a smoke screen. If he has sold only one or two of the boats you are interested in, then we'd suggest asking for names of people who have bought other boats from him. If we were still unable to get a list of a dozen or more people (satisfied customers), we'd seriously consider finding another dealer even if it meant going farther from home to do so.

Once you have a list, you can start with the assumption that the dealer will not give you the names of any customers he knows are dissatisfied, but that's O.K. Every dealer will have some dissatisfied customers. It's not possible to please all of the people all of the time. The basic question you want answered is whether the people you talk to are satisfied with the dealer's performance. This may include discussion of the dealer's handling of warranty claims, the work required in commissioning the boat, the manner in which money was handled and promises were kept. Perhaps most important, when the time comes to buy another boat, would they go back to the same dealer?

Financial References and Escrow Accounts

You are going to entrust the dealer and the boatbuilder with anywhere from $5,000 to $200,000 or more when you finally order your boat. Your interest here is in how your money will be handled. One glaring fact is that almost every year, some boatbuilders go out of business. Many of them have been successful as boatbuilders, but not as businessmen. In periods of inflation, builders may get caught by writing fixed-price contracts months before the boat will actually be constructed. As costs increase with inflation, the builder can end up losing money on each boat he turns out because his contract won't allow adjustment for the higher costs. When financial problems arise, it is not unusual for people who have ordered boats to be left holding the bag.

To help avoid these problems, we'd like the dealer to provide us with financial references. Banks may not tell us much, but the dealer's suppliers might. In any case, we'd ask for those references and a letter authorizing the references to answer our questions so long as the answers do not require specific financial information. All you really want to know from the suppliers is whether the dealer (or builder) pays his bills on time and reliably; you'd like the bank to reassure you of his financial stability and soundness. If the dealer won't provide those names and the letter, we'd again seriously consider going to another dealer. We'd also like assurances that our money would be kept apart from the dealer's and/or builder's operating funds by being placed in a separate escrow account. The builder would receive progress payments as work on the boat went forward and the purchaser would receive a regular accounting for those funds.

Warranty and Dealer's Responsibility

One of the most difficult areas is that of boat warranties and the dealer's responsibility toward that warranty. Our only advice in this area is to get everything in writing and to confirm the details with both the dealer and the builder. Along with the details of the warranty responsibility, you should also have a detailed set of specifications that the builder warrants will apply to your boat and a description of the standards of workmanship which your boat will meet when it is turned over to you. Only if you have these do you have a standard against which to compare the boat the builder sends to you. If the boat meets those specifications and standards of workmanship, you've bought it. If, however, there are serious discrepancies between the boat and those specifications and standards, you have leverage to negotiate correction of deficiencies. In the event of serious discrepancies that cannot be resolved in an acceptable manner, those specifications and standards may provide the insurance you need to cancel the purchase agreement.

9

Visiting the Factory
A Final Test

If you are at all serious about a specific boat, you should visit the factory before going any further in the decision-making process. This visit will serve two purposes: it will let you take a measure of the builder; and, it will provide opportunity for a more detailed look at the boat's construction to settle any questions you may have and to confirm through your own experience what you've been told by salesmen, brochures, owners, and reviewers.

In making plans to visit the factory, however, we suggest you give yourself some leverage by using the power of a purchase contract in your favor. Working with your lawyer, write a contract on the boat contingent upon at least two things: (1) your approval after visiting the factory; and (2) agreement upon a revised contract to be drawn after your return from the factory—assuming you still want to go ahead with the project. If you have not yet had a demonstration sail, you are well advised to add approval after a demonstration sail as a third contingency. By all means, have a lawyer look at the contract before you sign it to be certain the words mean what you intend them to mean. Your goal is this: You will have to put down a thousand dollars or more as "earnest money" when you sign the contract. You want to be sure that you can get your money back promptly if you want to cancel the deal.

How does a contract give you leverage? If you go to the factory as one who is thinking about buying one of their boats, you will receive courteous treatment, but probably not a major investment of time from the builder. On the other hand, if you go there as a live customer who has already signed a contract and put up some cash, the builder is going to bend over backwards to keep you on the hook. If, for example, people previously have been reluctant to provide much detail about the lay-up schedule, they now have a choice between giving it to you or having you cancel the contract. Preferably, you'd have details of the lay-up schedule before going to the factory so that you can compare what you're told out on the floor with what you've been told in the sales office. The two should match up.

When you get to the factory, expect to see a fairly well organized dog-and-pony show. All of these builders have taken many potential customers through their plants. They know that most customers not only don't know what to expect but also won't know what they're looking at anyway. They also know what they want the customer to see. They know what questions to expect, and how to answer those questions.

But you have an advantage. You also know what to expect. They expect you to be ignorant, and you certainly wouldn't want to disappoint them. They expect you to not know what you're looking at out in the plant, so you have license to ask virtually any curious question you want to ask. And ask you should. You want to be friendly, curious almost to a fault, admiring (whether or not you are), and always sympathetic to the builder's assertion that he's building the best boat on the market.

When you visit the factory, you should be dressed casually, but in clothes that you don't mind getting dirty. Men and women alike should wear long pants. Factory floors are not appropriate places for shorts; nor are skirts generally safe garments on a factory floor.

When you walk into the plant, you should have your objectives for the visit firmly in mind. We would suggest the following as a minimum: (1) measuring the health of the company; (2) assessing the quality of the building operation; (3) assessing the effectiveness of the quality control function; and (4) looking beneath the surface of the boat to examine details of construction that couldn't be seen in the finished boat.

MEASURING THE BUILDER'S HEALTH

You can start with a very simply query: How busy are they? They'll answer the question in the best possible light. But you can also answer the question for yourself just by observing what's going on at the plant during your visit. That means, however, scheduling your trip for a day when the plant should be operating normally by avoiding holidays. Your visit also should be planned either in the early morning or early afternoon. Particularly in warmer climates, work may start very early in the day so that it is completed before the hottest afternoon hours. The result may be a 10:00 A.M. lunch break and midafternoon quitting time. While the builder may want to tour you during lunch break, you should ask to make your tour during working hours.

When you are touring the plant, you should be curious about all of the operations. If the plant builds more than one boat, you'll want to know what other boats they build and ask to see some of the others in progress. Among other reasons, you may not be able to see all of the details you want to see on the particular boat you are considering. Looking at other models under construction may let you see how they execute similar details on those boats and you can ask whether they are carried out the same way on "your" boat.

As a businessman or woman, or even as a nonbusiness person who is merely curious, you might also like to know how many boats they can turn out a year. At some other time, you might be curious about how many boats they've built in the past couple of months. A comment about imported boats making it tough for U.S. builders because of lower wage rates in Far Eastern countries, or because of the value of the dollar against other currencies, might elicit a comment about business conditions. All of this intelligence helps you develop a picture of the economic health and viability of the company. And the picture can't be too clear. For example, we have one acquaintance who ordered a boat from one of the older fiber glass boat companies only to be asked for a progress payment before they had even started his hull lay-up. When he questioned the request, he learned that the builder needed that progress payment to meet his payroll. Other friends lost their half-completed boat when the builder suddenly filed for bankruptcy.

ASSESSING THE QUALITY OF THE BUILDER'S OPERATIONS

The quality of a finished boat reflects in part the care that goes into its construction. In our view, one of the clearest indicators of quality is the appearance of the factory. The laminating areas in some boat factories are incredibly messy. In others, you'd swear the workers could go about their jobs in their Sunday best. The difference is in the attitude of management and the pride of the workers. We believe it is likely that a management and work force that care enough to maintain a neat workplace will also care enough about each boat they build to pay attention to detail.

Another key indicator of quality is the stability of the work force. At some boat plants, worker turnover—particularly in the laminating shops—is frighteningly fast. In such a circumstance, it is difficult to have any confidence that the people who lay up your hull and deck will do the job right. Sometimes, idle questions about the difficulty of maintaining a stable work force will bring the information you want. You can also get some idea by talking to the workers themselves. For example: "That looks like a difficult job. How long did it take you to learn to do that?" As a follow-up question, if necessary: "How long have you been working here at XYZ?" And, "Is that when all of you (in this work group) started here, too?" Or (to another worker), "What about your job? What do you do? How long did it take you to learn to do that?" It is also quite natural to start your guide talking about his background, querying how long he has worked with that company and in that job. Nearly everyone enjoys feeling that he is important and that you are interested in him or her.

The working conditions provided by management can also affect boat quality. For example, do workers have to climb up and down ladders (or steps) to take work to their power tools? We know few if any people who don't get tired of going up and down steps or ladders. In a boat factory, such fatigue may provide strong temptation to let the pieces (parts of a boat) lie where they fall rather than go back to the power tools to make them fit correctly, particularly if the tools needed are off in another bay. Trekking back and forth and up and down is also inefficient, and adds to cost. Some builders, of course, recognize these problems and provide a raised work platform along the sheerline of hulls they're building. Required power tools are located on the platform so that a worker needing a table saw, for example, must only climb out of the hull onto the platform. Again, you are looking for attention to detail and planning, and for management efforts to provide good working conditions for its employees. All have a bearing on the quality of the operation and the workmanship that goes into each boat.

ASSESSING THE QUALITY OF QUALITY CONTROL

In planning your trip, we suggest that you try to schedule a meeting with the builder's quality control manager, to learn how he or she perceives his role. You want to learn two things: What kind of support he receives in his job from management; and, how his quality control system is organized. This means finding out where the quality control check points are in the production process, how they are monitored, what kinds of problems he tends to find, how they are corrected, and who settles disputes. We are happiest with frequent check points because problems invariably are easier to correct when they are discovered early. We are least happy with only a final quality check. As you know well by this time, it is simply not possible to make a thorough check of many critical areas after the boat has been completed. As a result, corrections that are made at the end of the production line usually are mostly cosmetic. If, on your visit, the quality control person is not available, you should still seek the same information, but you should also be given a pretty good reason for not being able to meet with him.

Ideally, the builder would have an independent person whose sole responsibility is quality control—making sure the boat is constructed according to specifications in a workmanlike manner—reporting either to the top management or, as a second choice, to marketing. The third choice is a quality control person who reports to manufacturing. The least desirable situation is placing responsibility for quality control in the foreman's hands.

Our bias against asking the foreman to handle quality control is that every problem he finds is a finger pointed at himself and his effectiveness in supervising the work force. The problem with housing a separate quality control supervisor in the manufacturing organization is that manufacturing's job is to produce the boats within defined cost limits. As a result, cost containment usually has a higher priority than quality control. Marketing, on the other hand, hears the complaints from the dealers and boat buyers. Marketing also pays warranty claims which arise during commissioning. Their objective is to market a product which makes dealers and boat buyers happy and keeps warranty claims at a minimum. Because they can cite chapter and verse, marketing is often effective in persuading manufacturing to accommodate their needs. Top management, from still a fourth perspective, has a broader view of the value of a reputation for quality products and of the costs of quality defects. It also has the clout which the quality control person may sometimes need to install and enforce effective quality control procedures.

LOOKING BENEATH THE SURFACE

You will have developed your own list of construction details that you want to see while checking out the completed boat. Essentially, you should be looking at construction details that can't really be seen by looking at the finished boat. You should also be keenly interested in the hull and deck laminations, including the lay-up schedule and procedures for doing the lay-up. Ideally, you could climb down into the hull and look closely at the laminate to search for milky patches (resin-starved areas) in the lay-up. In a cored construction, you'll also want to know what efforts the crew uses to be certain there is a good bond between the fiber glass materials and any core material. The latter is particularly important for either balsa or plywood cores, and this is one place where curious questions asked of the workers may be more revealing than questions asked of your tour guide. The guide probably knows what the answer should be; the workers probably know only what they actually do.

If the builder is active, he may have two or three boats of the model you're considering in production. With them all at different stages of completion, you have good opportunity to look at details. Often, however, only one boat will be in production. In that case, you need to schedule your trip when the interior is about one-half installed. If you go much later, you may not be able to see much more detail than you can see in the finished boat. Much earlier and you may be too early to see how through-hull fittings are installed, wiring and plumbing are run, and the liner and / or bulkheads are installed.

Any holes in the hull or deck laminate which you can examine are bonuses. They help you see the laminate thickness and, possibly, the coring firsthand. The edges of the holes should be checked for voids in the laminate or between the core and laminate. Looking at those holes also provides opportunity to ask your guide about the laminate schedule so that you can compare his answer to the other information you have received. We would also ask for detail about both the specific location and amounts of added reinforcement to support the keel and shroud loads on the hull and the designed overlap for the pieces of fabric in each laminate. That overlap should be a minimum of three to four inches. The builder may tell you (as more than one has told us), "We don't give out that information because it's meaningless unless you're a structural engineer." To that we can only say, "hogwash,'" and keep asking for the information.

This is also a good time to ask for one of the plugs removed in cutting the holes. Often they will give you the plug if they can find it. If so, you should find out exactly where it came from in the hull. Depending upon your level of confidence in the builder and the use intended for your boat, you may want to take the plug home to have it burned. In that way you can compare the laminate of the plug to the lay-up schedule you've been given. An extra layer of fabric may mean the plug came from an area where adjacent pieces of fabric were overlapped; fewer layers of fabric than expected would raise a question you'd probably like to have answered. You might have misunderstood the lay-up schedule, but they might have omitted a layer. It is unlikely that a builder would concede that he left out a layer of fabric, but he'll certainly need to explain the lay-up schedule in detail and the reason for the discrepancy if you feel that one still exists. Thereafter, you need to draw your own conclusions. If you think we make too much of knowing about the hull laminate, consider this: In the late 1970s, the hull of one popular 42-foot "cruising" ketch had the same lay-up schedule used for reinforcement in it as the icebox liner of a Westsail 32. Essentially the same boat is built today, but we are not privy to the current lamination schedule.

If the builder looks healthy, and all other signs are positive, we suggest going one more step. That is, asking whether the builder would allow you to hire a surveyor to check progress of your boat if you proceed with the contract. We'd suggest you ask the question whether or not you plan to actually hire a surveyor to do the work. The builder's reaction to the question is what you are interested in. If the answer is "yes," you can put it into the revised contract. If the answer is "no," you deserve a good reason for the refusal. If the problem is one of cost to the builder because he must have someone go with the surveyor every time he comes to inspect progress, then you can explore the possibility of reimbursing any reasonable costs incurred if the builder will tell you up front what those costs would be. If the charges are ridiculous, that tells you the builder really doesn't want your surveyor in his plant. At that point, whether that negative response to the surveyor query is enough to make you look at another boat would depend upon the use planned for the boat. Short of going out onto the ocean, you could probably live with the builder's refusal to have a surveyor inspect progress if you feel confident on the basis of all of your work to this point that the boat is suited for the sailing you'll be doing. For coastal or offshore cruising, however, we think you'll want a builder who is willing to open his shop to such inspection— even if you choose not to actually hire the surveyor for that purpose.

10

Making Your Choice
A Job Done Well

At some point, you will be ready to make your decision. It's an exciting moment. It is also the moment at which more than ever you need to operate from a basis of reason. One of the most important facts to keep in front of you at all times is this: you're in a buyer's market. In the first half of the 1980s, for example, there were only about 4,100 new auxiliary power sailboats sold each year in the United States and many of them were imports. At the same time, more than 300 builders—including many from overseas—were showing boats at the fall sailboat shows. Simple arithmetic suggests that each builder is selling an average of fewer than 14 boats a year. In actual practice, a small number of companies sell many more than that; a large number of companies, however, sell fewer than 14 boats each year. The story has been similar in the used sailboat market: There have been many more boats than buyers, with the result that most boats are on the market for several months or longer before they find a buyer.

What all of these statistics mean is this: You, as a serious boat buyer, are a relatively scarce commodity. As a result, *as long as you maintain control over the process by keeping your emotional attraction to the boat you're considering out of sight until after the contracts have been signed by all parties,* you're sitting in the catbird's seat.

PUTTING IT ALL TOGETHER

As you approach your decision of which boat to zero in on, it is well worth the time and effort to pause for a few days to go back to the beginning of this process, when you took the time to think seriously about the kind of sailing you'll be doing, and where you'll be doing it. Next, you used the results of that exercise to help figure out what category of boat you want to buy: one for dockside living; for sailing in protected waters; sailing in semiprotected water; coastal cruising; or for the open ocean. At the same time, you looked honestly at your capabilities and your boating needs to zero in on the size range you wanted to consider. Subsequently, you began filtering boats through a screening process designed to help you identify boats which appeared suited for you, your capabilities, your needs, and for the waters in which you will be sailing. And finally, you narrowed the field to just one or possibly two boats by going back to look at them carefully a second time, by qualifying the dealer, and by visiting the factory to see firsthand how they are constructed *and* to take a measure of the builder.

In theory, you've done everything right and the boat should be as close to perfect for you as compromise will allow. In practice, the boat could be a mistake.

How? In the same way that you can plot a course from point "A" to point "B," steer the course, and still miss your landfall by miles. Possibly a mistake was made in the plotting, or in converting from the True or Magnetic course to a steering compass course. Possibly there were current and/or wind factors you didn't take into consideration that affected the course made good over the ground.

In choosing a boat—as in piloting—it's worth going back to be certain that errors of judgment haven't crept into your efforts to find the boat you want. Perhaps you under- or overestimated your sailing capabilities. Or your sailing ambitions. Or perhaps in looking intensely at boat after boat in the past months, you or your spouse have learned things that are causing second thoughts that neither one of you has voiced for fear of hurting your partner's feelings, causing some disappointment or, even of admitting you made a mistake. Or, possibly still, you are right on course and the finalist boat is indeed just what you want. The point is this: It is important before you buy this boat to review all of the decisions you've made till now to be sure that you are still on the right track. Most likely, you'll find that you're right on target. If, however, you find that in retrospect your needs are different than you had thought, or that somehow you were diverted to the wrong course, you have ample time to go back and recalculate that course so that you find the right boat before you've signed the contract, not afterwards.

This is also a time to weigh options. Some of those options will involve features or equipment you are considering, but they may also involve a difficult choice between two boats. If you live on either coast and are considering two boats—one built in the east and the other in the west—don't forget to take into consideration the cost of getting the boat to you. Shipping a boat in the 30- to 35-foot range from Florida to the Mid-Atlantic states at this writing costs about $2,500 to $3,000. Shipping the same boat from coast to coast is double that amount or more. For boats larger than 35 feet, the costs go up rapidly because of width and height restrictions on the highways. The question you need to answer is whether the boat from the far location is worth the extra cost of getting it shipped to you.

Sometimes, of course, it is possible to work a deal on the shipping. For example, if the builder has potential customers in your area but no boats there for them to see, he may be willing to negotiate an arrangement in which he'll pay for shipping your boat if you'll agree to show the boat to prospective customers for him over a set period of time. Or he may need a boat for a boat show in your area and be willing to pay for shipping in exchange for the opportunity to put your boat in the boat show. (If you do have the opportunity to have your boat in a boat show, you may well enjoy the experience. We've worked a number of shows and found it great fun—as long as we don't have to do more than one show in a season. If, however, you would worry about possible damage to your boat from having a couple of thousand people—including children—come aboard in the four or five days of the show, don't do it. The heartburn isn't worth it.)

WHEN AND WHERE TO BUY

Sailboats of the type you're considering represent a seasonal market. Throughout most of the country, prospective boat buyers start thinking about sailing and looking for boats in the spring and summer months. In Florida, the season is the opposite. Interest in sailing peaks in the fall and winter months, and eases during the spring and summer. This seasonal marketing pattern has very practical implications for you as a boat buyer. If you're interested in buying a used boat, for example, the best times to negotiate, from your viewpoint, are autumn in cooler climates and late spring or early summer in Florida. In the north, sellers would rather not pay hauling and winter storage, and they know their boat is unlikely to sell in the winter months. In Florida, where summer months yield slow sales, it is costs of summer storage and bottom maintenance that are of concern.

The pattern is similar if you're interested in a new boat. The industry has developed a series of in-the-water and indoor boat shows on all three coasts scheduled to bolster the otherwise weak selling season. For a number of reasons, the major in-the-water shows represent particularly good times to close a deal. Many builders rely on sales from the boat shows to keep their production lines operating through the slow winter period and offer special "boat show packages" to entice show visitors to sign a contract. Moreover, it is expensive for dealers or builders to put their boats into a boat show and they want to get some return for that investment. Finally, the major shows are usually attended by boat company presidents, vice presidents, and national sales managers. These not only are people who have the authority to make price decisions, they also know where the year's business stands opposite their sales forecast, and they want to make that forecast—something they'll do only by selling boats. As if that weren't enough motivation, these management people also know both the status of their production schedule for the next six months and the importance of keeping their plant running. If the order book is empty, or nearly so, they are going to make concessions to try to keep the production lines operating.

Local dealers also have incentive to sell their stock at the boat shows. In states with year-end inventory taxes, dealers who have boats at the show may want to sell off their stock to reduce the burden of inventory taxes. Dealers also know that any boat not sold at the show probably will not be sold until spring. If they are financing their display boats, they're anxious to get out from under those interest charges.

CLOSING THE DEAL

When you go in to negotiate your purchase, your success in getting the best possible deal will depend upon your preparation for those negotiations and your ability to keep control of the situation. Preparation means developing a detailed list of the points you want to cover in your discussions. This involves reviewing all details of the boat, including specifications. If you want any options or changes in the interior, you should have those clearly identified. One of your objectives should be to obtain firm cost commitments for all elements in the contract so that you know before you sign the contract exactly how much this venture is going to cost you. You should be prepared to discuss payment schedules; you may be able to negotiate making your own progress payments upon receiving statements from the builder certifying the work that has been accomplished. If delivery time is important—for business reasons, for example—you might be prepared to discuss a performance clause in which you would be compensated by a modest reduction in price if the boat is not delivered as promised.

If you want to employ a surveyor to oversee progress on your boat during construction to ensure that it meets specifications of materials and workmanship, this is the time to negotiate agreement for you to use such a service. Alternatively, if you want to have the contract contingent upon an acceptable marine survey of your finished boat, this is the time to negotiate that detail. Obviously, if you do use the services of a surveyor, the payment schedule must be arranged so that you retain some financial leverage throughout the building and delivery process. Your attorney can help you with that detail.

If you are going to use a surveyor, it is important that you be realistic about what he can do for you. If he is inspecting progress during construction, he can assure you that the boat is constructed to specification and to good standards of workmanship. If he finds a problem, he may be able to get it resolved simply by pointing it out to the builder. If not, however, all he can do is keep you informed so that you can negotiate with the builder to have the problem corrected. But it's up to you to get the problem solved.

If you hire a surveyor to go over your boat when it is delivered, all he can do is provide you with a thorough assessment of how the boat stacks up against the specifications and the standards of workmanship agreed upon in your purchase contract. If your surveyor finds a number of small items which he believes should be corrected before you accept delivery of the boat, you can begin negotiating. If the surveyor, after discussing your sailing plans with you, makes suggestions for enhancing the boat's suitability for the use you intend, those items are not negotiable with the dealer before delivery. Rather, they are opportunities for you to spend extra money in improving your boat—if you agree with his recommendations, which you may not. If the boat is grossly out of whack, the surveyor will probably discover that fact and you at least can begin negotiating before you've made the final payment and accepted delivery of the boat. In some cases the boat may be so poorly done that you must refuse it. We know of one recent instance in which an $87,000 sailboat was hauled back to the factory in Florida from the Chesapeake Bay area when the survey turned up so many voids in the hull laminate that the boat could not be repaired satisfactorily. The buyer was sent a replacement boat which also had a number of voids in the lay-up, but not so many they could not be repaired in commissioning.

When it comes to actually negotiating a price after all of the details have been agreed upon, you must be prepared to walk away from the deal if you are not satisfied with the offer you are receiving. It's a question of who blinks first, or one of who has the strongest need to complete the deal. But you also need to be realistic. You should already know the standard list price of the boat and the cost of any options you are considering before you get down to negotiating the final price. That knowledge helps you figure out how much you are willing to pay. In that way, you can negotiate toward a specific price; otherwise you're just trying to see how low you can make the dealer or builder go. If you get him too low, he may enter the deal unhappy and, if that happens, you could wind up losing instead of winning. If, on the other hand, you have a price in mind that you believe is realistic, you have a better chance of negotiating a deal that both of you are happy with. When both of you are happy, you've made a good deal.

If you are negotiating with a dealer, he normally has only a 17 to 20 percent spread on the base boat between his cost and list price. That spread between cost and list price has to pay its share of the dealer's overhead, the salesman's commission, and still provide some degree of profit after reaching a price agreement with you.

The dealer generally has a much higher margin on optional equipment. You can get some idea of the dealer's costs by pricing items in discount catalogues. For most items, the dealer pays somewhat less than the discount price in the catalog. If the dealer will give you a price almost competitive with the discount catalogues, it may be worth purchasing the items through him—particularly if his price includes installation and he can service that equipment if it malfunctions.

There are other ways for dealers to improve their offerings as well. For example, some dealers will purchase a few genoas for popular boat models at a very low price from local sailmakers during the off season when the sailmakers are hungry. The sails may not be of the best quality, but they let the dealer offer his customer a genoa to sweeten the package.

In negotiating with a builder, you may have somewhat greater latitude because you've got his profit margin and, sometimes, the dealer's margin to work with. The builder's costs for optional equipment are lower still than the dealer's costs. However, you need to be certain that whatever price you negotiate takes into consideration the cost of commissioning your boat when it is delivered. That cost is one that the dealer normally absorbs. If the boat is in good shape as it comes off the truck, the cost will not be excessive. If, however, there are quality defects from the factory—as is too often the case—repairing those defects can cost hundreds or even thousands of dollars. If commissioning is not included in the price, you should get estimates from at least two yards who have experience in commissioning boats from your builder so that you can take that into consideration in your negotiations and decision.

When all negotiations have been completed, there is still one final step we would urge upon you. In boat selling, much is done verbally and with a handshake. In theory, that's a nice way to do business. In practice, particularly with the amount of money you're spending, it's a practice fraught with difficulty. The problem is not usually one of honesty or intent. Most often, it's a problem of communication. The buyer thinks he's been promised one thing and is surprised when he gets something else. So it is in both your own self-interest and the dealer's to have everything put in writing—every detail which you consider at all important—and then checked by your attorney to be certain the written words say what you intended them to say. When that is done, you can sign the contract with confidence. And then let your emotions run. Your dream is on its way to reality.

Appendix

Yacht Brokers
Making Them Work for You

Buying a brokerage boat ought to be a straightforward and pleasant experience. In fact, most of the time it is. Yacht brokerage, however, is a complex business and tripping over some of those complexities can not only sour your experience, it can cost you money. There is a way to avoid problems and that is to take the business just as seriously as you take the job of choosing your boat. From a practical standpoint, this means learning enough about the yacht brokerage business to use the system to your advantage rather than allowing yourself to be a passive participant and—possibly—a victim.

On the surface, yacht brokering hasn't changed much over the years. Yacht brokers today have the same basic functions they have always had: (1) Selling previously owned boats; (2) helping people buy previously owned boats; and (3) helping arrange for the contribution of boats to charitable institutions. Scratch the surface, however, and the changes leap out at you.

• In 1960, there were about 20 yacht brokerage firms in the two-county area that includes Miami and Fort Lauderdale. Today, the Fort Lauderdale telephone book alone lists some 200 brokers. The number of brokerage offices nationwide is anyone's guess.

• In 1960, the majority of boats up to 60 or 70 feet were constructed of wood. Today, most boats in this size range are produced on a quasi-assembly-line basis of fiber glass or, in larger sizes, of aluminum or steel.

- In 1960, most boat owners started with small boats and traded up through several boats to larger sizes. Today it is common for people to start out with a 40-footer or larger.
- In 1960, most boats were cash purchases. Today, boat mortgages are commonplace and they often run as long as 15 years.

The practical effect of these changes has been the development of a very large market and that in turn has made yacht brokering an attractive business. Moreover, in the words of more than one broker, "everybody wants a piece of it."

AN INDUSTRY IN TRANSITION

The rapid growth of the industry has brought many changes in the business. At the trailing edge of this change is what might be called the "traditional" view of the business. At the leading edge is a view that sees yacht brokering principally as a way to turn a profit. Although neither view has a monopoly on quality or sophistication of service, each view does color the broker's approach to the business.

Traditionally, people have gotten into yacht brokering because they have loved boats and boating. It's been a way to combine avocation and vocation—a fact which has often led to a close sense of identification between the broker and his clients. "We have clients, not customers," says one such broker. "There's a lot of sincere counseling. I don't want to sell a boat that is unfit either for the use intended or for the people buying it." The view from the other end of the spectrum is that the business is "too romantic." "There's a sad lack of good professional salesmen," one of the less tradition-oriented brokers says. "There's a feeling that all you need to be a good broker is to have spent your life around boats. That isn't true. Get me a good, hungry salesman and I'll teach him the pointy end from the blunt end." Most brokers, of course, fall somewhere between the two extremes.

STRUCTURE OF THE INDUSTRY

As the yachting industry has grown, the yacht brokerage business has tended also to become more specialized. One result has been a layering of the business by boat price. In general, there are four layers:

• Large firms with offices in four or five major boating centers in the U.S. and, possibly, Europe. These firms handle mainly large yachts ranging in price from $200,000 up.

• Midsize firms with offices in only one or two major boating centers, whose business falls principally in the $75,000 to $200,000 price range.

• Smaller firms with just a few salesmen and one location. Most of their business is in the $25,000 to $75,000 price range.

• Local brokers, often associated with a marina, whose business is mostly in the under $25,000 range.

As the buyer, your chances of getting the service you need are greater if you select a broker whose selling interests are in your price range. The key, however, is to make sure you fit into the middle-to-upper price range of the boats your broker usually handles. The reasoning is simple: If a broker has listings which range from $25,000 to $75,000, more effort will go into selling the higher-priced boats because 10 percent (selling commission) of $75,000 is three times 10 percent of $25,000.

The size of the commission, however, is not the only reason brokers prefer to work with more expensive boats. There is a conventional wisdom that people who are looking for a $50,000 boat are likely to be more sophisticated in the ways of business than a person looking for a $15,000 boat—and, therefore, easier to work with. In addition, some brokers say that a person looking for a lower-priced boat requires more time and effort than the sale is worth—in part, as one broker said, "because he's trying to get $20,000 worth of boat for $15,000." Conventional wisdom also says that an expensive boat is usually well maintained and will have little trouble passing a survey, whereas a lower-priced boat is likely to be less well maintained and have more trouble with a survey.

The practical result is that many brokers have placed lower limits on the price of boats they will handle—for example, $25,000 or $100,000. If a broker has such a limit and the boat you want to buy just skims in over it, you might be well advised to find a broker with a lower limit. You want to find a broker who will consider you among his more promising prospects.

HOW THE SYSTEM WORKS

Yacht brokering begins with a boat listing—the agreement between a seller and a broker which authorizes the broker to try to sell that boat. There are three ways to list boats with a broker: central, exclusive, or open. Most brokers will encourage owners to give them a "central listing" and generally it is to the owner's advantage to do so. With a central listing, the broker can spread the effort to sell the boat to other brokers, splitting the commission with them if they sell the boat. This increases the boat's exposure to the market, thereby improving the likelihood that it will be sold expeditiously. From your viewpoint as a buyer, it is also this which enables the broker you choose to show you most any boat you want to see. Occasionally, a broker may want an exclusive when the boat being listed has great market appeal, but that exclusive usually will limit the boat's exposure to the market and may delay its sale. An "open listing" is principally a way for an owner to list his boat with a number of brokers while retaining the right to sell it himself without a commission.

COMMISSIONS

A commission of 10 percent of the selling price of the boat is standard throughout the industry regardless of the kind of listing agreement. With central listings, the usual practice is for the listing broker to receive 30 percent of the commission, and the selling broker 70 percent.

SELECTING A BROKER

The proliferation of boat types has caused increased specialization in the industry. It is virtually impossible today for any one broker to be expert on all types of boats. As a result, it is important for you as the buyer to find a broker whose specialized knowledge (if he or she is an experienced boat person) encompasses your boating interests. As important as it is to land a broker qualified to help you, it can be difficult unless you know someone or are willing to assert your needs. The cause of this difficulty, aside from the proliferation of broker salesmen who barely know the "pointy end from the blunt end," is known as "floor duty." Yacht brokers—like their land-bound counterparts, real-estate brokers—assign their salesmen to floor duty on a rotating schedule so that the office is covered at all times. Prospective clients who come into the office and do not ask for a specific salesman by name are assigned to the person on floor duty that day. If you are interested in a cruising boat and your salesman is a hot-shot SORC racer or powerboat enthusiast, he is less likely to be helpful to you than if his personal interest and experience is with cruising sailboats.

One way to learn whether the salesman is qualified to handle your specific needs is by interviewing him (Chapter 4). Get him to talk about his interests and experience. Another is to look through his listings (the boats for which he is the listing agent) and the record of boats he personally has sold in the last year. If the boats you are interested in would fit right into both of those groups of boats, you may have found a good broker for your needs. If not, you should probably consider looking elsewhere.

There are also a number of business-related factors you need to consider in selecting a broker—factors which may have little to do with the kind of boat or price range the broker handles, but much to do with how he conducts his business. Sometimes you can learn about a broker by looking into the membership requirements of associations he belongs to. For example, to become a member of the Southern Yacht Brokers Association (SYBA) in Florida, a broker must have been in business for at least two years, be bonded, use an escrow account for client funds, submit letters of recommendation from four SYBA members, and list the owners of his brokerage business. It is, therefore, a reasonable assumption that any broker who is a member of the SYBA is reputable. (In looking into such membership requirements, however, ask the associations themselves, not the broker.)

Unfortunately, most brokers don't belong to such associations. Or the membership requirements of their associations are considerably less stringent than those of the SYBA. Because of this, you may need to consider other factors in attempting to make sure that you're dealing with a going concern. Even a small broker, for example, should have his office organized so that it is always covered. From a boat seller's viewpoint the reason for having the office covered at all times is obvious: his boat can't be sold if there's no one there to answer the telephone or to greet you when you call. From your viewpoint as the boat buyer, the need for having the office covered is no less important. Finding a broker who has an adequately staffed office is a prerequisite to your sense of security when handing him a cashier's check or arranging a wire transfer of thousands of dollars into his account—even if it is an escrow account.

You should also consider seeking business references from a broker. For example, you might ask about his list of corresponding brokers—the other brokerage firms with whom he has arrangements for sharing listings. You want to know who they are, how many there are, and, most importantly, which ones he has actually had business dealings with by selling one another's listings. You might also ask for the names and addresses of other clients he has helped buy a boat like those you are considering. Then check those references. A reference left unchecked is worthless.

If you are a serious boat buyer, the broker should be willing to spend time with you to learn about you, your boating experience, and your boating aspirations. In addition, he should be willing to take you aboard several boats early on—not in an effort to sell you those boats, but to see how you react to different features about them. This is one way he has to learn enough about you to help you focus on boats which are most likely to be of interest, rather than merely taking you to look at every boat in your size and price range. Of course, if you have already done the preselecting and developed your own list of candidate boats, you can save him and yourself time by telling him about the boats you are considering and why.

If you find a broker who is willing to invest his time and personal attention in you—assuming that you have qualified him to be your broker—you owe it to him to stick with him until you select your boat. And then, be sure you let him sell the boat to you. Unless you have negotiated a separate fee arrangement with him, the only way a broker is paid for helping you, often educating and counseling you along the way, is by selling you the boat you ultimately buy.

WHOM DOES THE BROKER SERVE?

Once a yacht has been listed with a broker, the focus of the yacht brokerage system shifts to you—the buyer. It's a shift that raises what some brokers find a sensitive question: "Whose interest does the broker serve?"

The basic fact, of course, is that the broker, in the words of one, "is working to put a deal together." That's how he gets paid. However, except in the unusual circumstance in which a buyer pays a finder's fee, the broker has a specific obligation to the seller, since that is who pays his commission. In practice, brokers have to walk a tightrope in attempting to serve the interests of both parties. One broker, for example, says, "Legally I'm working for the seller. But any broker worth his salt works for both parties as much as he can." Another is more explicit: "I'm serving the interest of the buyer to help him select a boat for his needs, but once he zooms in on a particular boat, I'm serving the financial interests of the seller."

A finder's fee offers you as the buyer a means of ensuring that your broker works in your interests only. Essentially, you and your broker agree upon a fixed fee to be paid in lieu of a commission for his services. When you purchase your boat through that broker, the commission he would normally receive for selling the boat is negotiated out of the purchase price. The way that the brokerage system works, however, a buyer usually does not have to offer a finder's fee in order to get the kind of service he wants from a broker. Instead, the key is as simple as (1) selecting a broker you are comfortable with and who takes an interest in you, and (2) sticking with that broker right on through the completed purchase of your boat.

CONTRACTS AND DEPOSITS

Once you select the boat you want, the next step is to make a purchase offer. This involves signing a purchase agreement and giving the broker a deposit equal to 10 percent of the offered price. Purchase contracts vary greatly. Some are filled with language that means something only to lawyers, while others are simple and straightforward. Whatever the language and format, however, you should be certain the contract is contingent not only upon agreement on price, but also upon four other items designed to protect your interests:
• Completion of an acceptable survey;
• Completion of an acceptable sea trial;
• Completion of any financial arrangements necessary for you to carry through with the purchase; and,
• Presentation by the seller of basic papers required for U.S. Coast Guard documentation (or redocumentation) of the yacht—even if you do not now intend to have the boat documented. In addition, the contract should require a written or telegraphed response from the seller to any purchase price offered, with the amount of your offer specifically noted in the response.

Marine Surveys

Most brokers will have the names of several surveyors qualified to survey the boat you are interested in purchasing. The broker should not limit his suggestions to only one or two in order to avoid any possible appearance of collusion between him and the surveyors. It is up to you, of course, to employ the surveyor, and any costs involved will be your responsibility.

A survey may report that the boat is in generally good shape and is well suited for the use intended. It may also note deficiencies the surveyor believes should be corrected before the boat can be operated safely. As the buyer, of course, you can terminate a purchase agreement and have your deposit returned if the surveyor finds such deficiencies. That's what the contingency clause is all about. However, you also have two other options: One is going ahead with the purchase despite the deficiencies; the other is renegotiating the purchase agreement to require the seller to correct deficiencies found in the survey before the sale goes forward. If the latter option is chosen, any repairs should be resurveyed before the purchase is completed to ensure that they were made properly.

Sea Trials

Most people wouldn't think of buying a car without test driving it. A sea trial is to a boat what a test drive is to a car. However, depending upon your boating experience and the kind of boat you are buying, you might consider hiring a licensed skipper to go with you as a consultant for the sea trial. In addition, a word of caution: There is a great temptation on a sea trial to sit back and enjoy the ride. There is also temptation to make excuses for a boat's poor performance. Don't fall into either trap. It is important to put the boat through a variety of paces to be reasonably sure the vessel's range of handling and operating characteristics will satisfy your needs. If weather conditions don't permit such testing on the day of your sea trial, go out another day. The time to discover that you are dissatisfied with the boat's performance is before you buy it, not afterwards.

Financial Arrangements

Many boats are purchased through some kind of boat financing arrangement. Your broker can often suggest local sources of financing, but you should shop for financing the same way you shop for anything else. Be aware also that your broker may receive a fee from some financing institutions for bringing you in as a customer. The fee is called a "dealer's reserve." Unless you ask, you won't know whether or not the broker is receiving it. Some brokers claim they pass the "reserve" along to the buyer, but don't count on it.

Coast Guard Documentation

Even if you do not plan to document your boat, you would do well to insist upon having the essential legal papers required for documentation as a condition for purchasing the boat. You should do this if for no other reason than the fact that the next buyer may want these papers from you. In addition, however, it could save you many problems if the boat you are considering was built in another country or was remodeled or repaired (other than emergency repairs) outside of the U.S. The reason is that an import duty should have been paid to U.S. Customs on any foreign-built boat brought into the United States, or on the cost of any repairs or remodeling done in another country. If that duty has not been paid—even several owners ago—the boat can be confiscated by the Customs Service. Although it is possible to petition the Customs Service to permit you to use the boat while the case is being investigated, you cannot sell the boat or transfer ownership while the case is pending.

The papers needed to document a foreign-built boat—in addition to those required for yachts constructed in the U.S.—are a Consumption Entry Form to show that duty has been paid, and a Certificate of Bona Fide to show that the boat's overseas title has been carried out in conformance with U.S. laws and is permanent. If work was done on a boat outside U.S. waters, that work should have been declared when the boat was returned to the U.S. If it was declared, the entry papers will show that the duty was paid. If the boat was imported directly by the original owner, the duty papers will show whether duty was paid. Many brokers are not aware of these requirements, so you need to be aware of them. Detailed information is available from the U.S. Coast Guard and from the Customs Service.

Written Response to an Offer

Although brokers customarily submit offers in the order in which they are received, it doesn't always happen that way. As a result, the only way you can be certain that your offer was submitted is to require written confirmation from the seller (letter or telegram) that your offer *with the amount noted* has been accepted or rejected.

Deposits

Most brokers will not accept your personal check for a deposit on a boat. It's nothing personal, just good business practice. It simply takes too long for personal checks to clear through the banking system. Most brokers, however, will accept a bank cashier's check. Many will accept a wire transfer of funds directly into an escrow account.

Until the broker has cleared funds on deposit, you should not expect him to arrange for your surveyor to survey the boat. Nor should you expect him to give you a sea trial before the deposit has cleared. In fact, some brokers—particularly for larger boats—will not even submit your offer until they have cleared funds on deposit. These precautions are intended to protect the seller's interests. For your protection as the buyer, ask your broker to place your deposit into an escrow account. For his part, the broker should give you a desposit slip showing that the money has indeed been escrowed.

Although your deposit represents the potential commission on the boat, the funds remain yours until the sale is consummated or unless you forfeit the deposit. In case of forfeit, the funds are sometimes divided equally between the seller and the broker. If a broker does not have an escrow account, you might suggest depositing the money with a mutually acceptable attorney to be held in escrow. If the broker is reluctant to do one or the other, you might consider finding another broker.

STORM WARNINGS

In general, of course, yacht brokers are like any other diverse group of people. Most are honest. Most do their best to do their jobs well. Problems can arise, however, in buying a boat and you need to recognize the storm warnings when they are flying.

For example, you should back off quickly if a broker suggests that you can stretch the limits of the contract he's encouraging you to sign. Or, if the broker dismisses the contingencies you want to include by implying that they may discourage the seller, that they're unnecessary, or that you can add them later, don't believe it. And be careful!

There are other warning signs as well:
- Beware the broker who wants to sell you only his own listings.
- Beware the broker who pressures you to buy a boat.
- Beware the small broker who insists on meeting you at a boat rather than at his office. You may wish to be certain that he has an established and properly staffed office.
- Beware the broker whose listing book—if he shows it to you—has all of the listings copied onto his own letterhead. That is one way to inflate the apparent size of his business.
- When going to settlement, make certain that both you and the seller are represented at the settlement proceedings and that all financial details of the transaction are on one piece of paper for all parties to see. In any complex and multi-staged transaction like the sale of a yacht, a complete summary will make it possible for the "civilians" to keep track of the snowdrift of necessary papers.

Of course, storm warnings in business dealings are like those at the marina: They fly only occasionally. Most of the time, buying your boat will go without major problems. However, by putting your knowledge of how the system works to good use in selecting a broker, you can do better than merely buying your boat without major problems. Instead, you can make the system work for you in a positive manner. In this way, you will still like your boat and be looking forward to a new season of sailing a year later, rather than to one of selling.